Biochemistry and Structure of Cell Organelles

TERTIARY LEVEL BIOLOGY

A series covering selected areas of biology at advanced
undergraduate level. While designed specifically for course
options at this level within Universities and
Polytechnics, the series will be of great value to
specialists and research workers in other fields who require
a knowledge of the essentials of a subject.

Titles in the series:

Biochemistry and Structure of Cell Organelles

ROBERT A. REID, B.Sc., Ph.D.

Lecturer in Biochemistry
Department of Biology
University of York

and

RACHEL M. LEECH, M.A., D.Phil.

Professor in Plant Sciences
Department of Biology
University of York

Blackie
Glasgow and London

Blackie & Son Limited
Bishopbriggs
Glasgow G64 2NZ

Furnival House
14–18 High Holborn
London WC1V 6BX

International Standard Book Numbers
Hardback 0 216 91004 8
Limp 0 216 91005 6

Filmset by Willmer Brothers Limited, Birkenhead
Printed and bound by Wm. Clowes (Beccles) Ltd., Beccles and London

Preface

THIS BOOK HAS BEEN WRITTEN BECAUSE WE FEEL THAT THERE IS A NEED FOR AN up-to-date compact book on cell organelles that transmits the excitement and challenge of modern subcellular biology. We hope that the book will be interesting and useful to students of the biological sciences and medicine, and to those in the teaching professions who do not have ready access to research papers.

Since space is at a premium, we have denied ourselves the luxury of a philosophical discussion of the problems of defining organelles. Rather we have chosen to include all those intracellular structures which have limiting membranes and definable compartments. The separate chapters consider nuclei, plastids, mitochondria, microbodies, endoplasmic and sarcoplasmic reticulum, Golgi bodies, lysosomes and various secretory vesicles, including chromaffin granules and synaptic vesicles. Nucleoli, ribosomes, and centrioles are included in the chapter on nuclei. New and exciting information about all these structures has emerged in recent years—for example, the nucleosome, interrupted genes, signal sequences on proteins destined for the bioenergetic organelles, mapping and sequencing of organelle genes, and consolidation of chemiosmosis as a unifying principle in energy transduction. We have outlined as many of these developments as possible and pointed out some areas of controversy.

The literature on subcellular biology is so extensive that it would have been easier to have written a separate book on each organelle. But that would have denied us the opportunity to discuss how organelles interact and complement each other in a myriad of diffent ways during the lifetime of the cell—an appreciation of which is central to an understanding of metabolism, growth and differentiation. How does a cell "know" that it has enough chloroplasts? How are events on the plasma membrane related to electron transport in mitochondria? How is the balance between exocytosis and endocytosis achieved? As yet there are few answers to such questions. But it is important to ask them, since questions suggest hypotheses, and hypotheses inspire experiments, and experiments eventually bring enlightenment.

R.A.R.
R.M.L.

Acknowledgments

It is a pleasure to acknowledge the constructive criticism and advice of many colleagues—particularly Dr. S. A. Boffey, Dr. S. J. S. Hardy, Dr. U. Hibner, Professor D. Robertson, Dr. M. Somlo, Mrs. L. Skiera and Dr. D. S. C. White. The book would not have been completed without the technical assistance of Mrs. M. Messer and Mrs. J. Donnelly. The figures were primarily the work of Mrs. Sue Sparrow.

We are indebted to the following colleagues for allowing us to use their original electron micrographs: Dr. J. Baker, Dr. M. S. C. Birbeck, Dr. G. R. Bullough, Dr. R. S. Decker, Professor W. W. Franke, Dr. A. D. Greenwood, Mr. P. G. Humpherson, Dr. B. E. Juniper, Dr. D. Lawson, Dr. D. N. Luck, Dr. G. G. Maul, Dr. J. J. Rose, Dr. W. W. Thomson, Dr. D. Whitehead and Dr. A. Wilson.

Where the line illustrations are based on previously published work, we have acknowledged the source under the figure or in the adjacent text.

Contents

CHAPTER ONE

CELL COMPARTMENTATION

THERE ARE MANY DIFFERENT CELL TYPES AMONG FUNGI, PROTOZOANS AND higher plants and animals. They differ in size, form and function, degree of specialization and mean generation time. Yet at the ultrastructural level there is a sameness about cells that is almost tedious. The same basic structures—nuclei, plastids, mitochondria, endoplasmic reticulum, Golgi bodies, lysosomes—appear with predictable regularity (figure 1.1). Whether this reflects a monophylogenetic origin of higher cells or convergence in response to selection pressures will be discussed later. The immediate point is that the monopoly position of these cell organelles in nature suggests that they comprise a winning formula, that is, a formula that facilitates the existence of cells as self-replicating self-regulating systems that operate according to known chemical and physical laws. However, there is another winning formula in nature, shown by bacteria and blue-green algae. These prokaryotic cells, as they are called, are immensely successful in terms of number of species and range of environments occupied. They carry out metabolic processes that are essentially similar to those of higher (eukaryotic) cells. They have sophisticated genetic systems, they synthesize complex macromolecules, they have efficient energy transforming mechanisms, and they show growth rates that are generally much faster than those of higher cells. Yet prokaryotic cells do not have internal organelle or membrane systems comparable to those in eukaryotes. Is a two-tier system of cell organization really necessary? What is the functional value of membrane-enclosed compartments in higher cells?

The most obvious example of compartmentation is the enclosure of most of the eukaryotic genome in a nuclear membrane, a feature that is probably related to the multiplicity of chromosomes in higher cells. Figure 1.2 shows some theoretical possibilities for the distribution of chromosomes in cells. Scheme A, in which the chromosomes are randomly distributed throughout the cell, would need an elaborate system for the precise segregation of daughter chromosomes during cell division. Scheme B

1

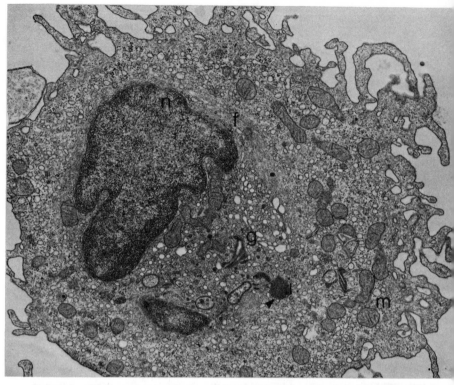

Figure 1.1a   Electron micrograph of a rabbit macrophage from lymph draining skin of hind limb, showing nucleus (n), mitochondria (m), Golgi bodies (g), lysosomes (arrowed) and myofibrils (f). (Dr. Gillian Bullock) (× 20 000).

shows one possibility of organizing the genome so that segregation occurs with reasonable precision. In this scheme the chromosomes are attached to the plasma membrane equatorially, and separation of the daughter chromosomes is brought about by membrane growth between the attachment points. This is similar to models proposed for the replication of the single chromosome in prokaryotes and separation of the daughters. The factors that have operated against its development in eukaryotic cells can only be speculated upon. In fact a position near the cell surface would seem less than ideal for genetic material, as the microenvironment there is probably more subject to fluctuations of solute concentration and pH than in the interior, and the genome would be particularly vulnerable to radiations and chemicals from outside. A more serious problem could be

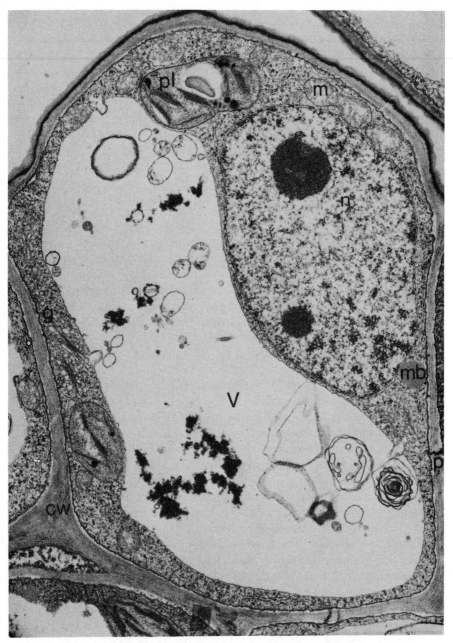

Figure 1.1b Electron micrograph of a young developing bladder cell in a leaf of *Atriplex* showing cell wall (cw), plastids (pl), nucleus (n), microbody (mb), vacuole (v), Golgi bodies (g), mitochondria (m) and plasmadesmata (p). (Dr. W. W. Thomson) (× 8000).

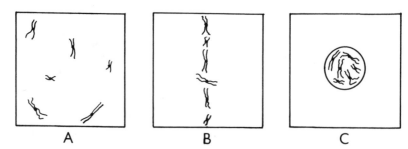

Figure 1.2   Theoretical possibilities for the organization of chromosomes in cells.

the large size of the eukaryotic chromosomes. Although they are never in their fully extended form (*c.* 70 cm for the longest human chromosome), a degree of open structure is necessary for replication and the transcription of DNA to RNA, and so they must be in an environment that allows transitions between these relatively extended forms and the tightly condensed states that are essential for the mechanics of segregation. As we shall argue later, higher cells require a system of canaliculi, vesicles and bioenergetic organelles, and it is difficult to see how these could exist in the same compartment as chromosomes without creating mechanical problems. An inner compartment for chromosomes (scheme C) such as that provided by a nuclear membrane has several advantages. The chromosomes can uncoil for replication and transcription, and supercoil for segregation without impedence. The genes are more protected than if they were randomly dispersed or associated with the peripheral membrane. The possibilities for segregation are less restricted than in the other schemes. In theory it might be argued that a specific compartment for genes provides the opportunity for independent control of the gene environment. In practice, the nucleus is unable to concentrate substrates and other low-molecular-weight materials against osmotic and electrical gradients, and a regulatory role for the membrane has not been clearly demonstrated.

All cells synthesize protein, the dogma being that the DNA of the genes is transcribed to RNA, which is translated to protein on nucleoprotein particles, ribosomes. In prokaryotes, these processes are not separated by a membrane, and it is not unusual for translation of an RNA molecule to be occurring at one end while the other end is still being transcribed from the DNA. In contrast, in eukaryotes protein synthesis occurs outside the gene compartment in the cytoplasm, despite the fact that both the RNA templates and the ribosomes are assembled in the nucleus. The most plausible explanation for the exclusion of protein synthesis from the nucleus is that a goulash of chromosomes, ribosome-studded RNA molecules, growing polypeptide chains and essential enzymes would be unmanageable unless the nucleus was larger than the optimal size dictated by other considerations; in other words, if the nucleus were large enough to handle efficiently protein synthesis as well as gene replication and transcription, some of the aforementioned advantages of having a nucleus at all might disappear. The intervention of a membrane between the processes might provide a means of regulating protein synthesis by, for example, influencing the rate at which RNA and ribosomes became available. However, there is no firm evidence of this.

Many eukaryotic cells synthesize proteins for export as well as internal use. Most prokaryotes also secrete proteins, e.g. for cell-wall synthesis or functions in the periplasmic space, and in these lower cells this is done in a novel way. In brief, proteins for export have "leader" sequences of amino acids that are able to permeate the plasma membrane and thread the remainder of the protein through as it is synthesized (figure 1.3). This close association between protein synthesis and transport through the membrane is made possible by special ribosome binding sites on the inner face of the membrane. The vital point is that, after the RNA-ribosome complex forms in the cytosol, it must bind to a membrane site before the growing polypeptide is more than about 20 amino acids long. Otherwise the polypeptide assumes a conformation that makes it impossible for the leader sequence to thread into the membrane. The deadline appears to be about 5 seconds after the RNA-ribosome association has formed. Therefore the prokaryotic model is only applicable to cells that have a sufficiently high surface area to volume ratio to ensure a high frequency of random collisions between loaded ribosomes and plasma membrane. Although this is clearly the case for prokaryotes, it is not necessarily so in larger cells with lower surface area to volume ratios. For instance, it would

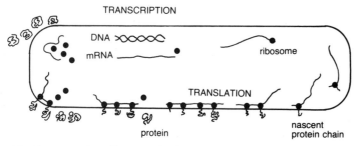

Figure 1.3 Representation of how proteins for secretion are synthesized in prokaryotic cells.

take a charged ribosome about 150 times longer to diffuse the 11 μm or so from the nuclear membrane to the plasma membrane of a liver cell than it would from the centre of a typical coccoid bacterium (1·4 μm diameter) to the periphery. Many eukaryotic cells are much larger than liver cells, and it would seem impossible for them to meet the spatio-temporal conditions under which prokaryotes operate. Another limitation of the prokaryotic system is that protein secretion is obligatorily linked to synthesis, and there is little scope for storing proteins and discharging them quickly in response to stimuli. Opportunities for post-translational modification of proteins for export are also limited. Taking account of the large size of eukaryotic cells and the fact that many of them secrete modified proteins in response to external stimuli, even if we knew nothing about their ultrastructure we would be forced to predict that they had a more elaborate transport and processing system than there is in prokaryotes. As discussed in chapter 6, this takes the form of a series of compartments in the endoplasm, a membrane-enclosed reticulum which, in the case of active secretory cells, provides a much greater surface area than does the plasma membrane, and receives RNA-charged ribosomes that thread proteins through into the lumen in a similar manner to that developed in prokaryotes. Once segregated into this endoplasmic reticulum, the proteins can be chemically modified at leisure and packaged into secretory vesicles that can be discharged at the plasma membrane in response to internal or external chemical signals.

It has long been known that seemingly incompatible processes occur in higher cells; for example, protozoans like *Amoeba* and *Peranema* degrade the carbohydrates, proteins, lipids and nucleic acids of the organisms they engulf to basic units, many of which are simultaneously synthesized back into carbohydrates, proteins, lipids and nucleic acids as growth proceeds. To a greater or lesser extent, the ability to degrade imported and even internally produced material is a feature of all animal cells and some plant cells. The degradation is brought about by a spectrum of hydrolytic enzymes that show optimal activity at acid pH and collectively are able to break down every biological polymer known. Historically, these acid hydrolases were postulated as being in cell compartments well away from susceptible cell material before this was actually shown to be the case by microscopy and separation of the organelles concerned (lysosomes). In containing a unique set of enzymes and facilitating processes that would be incompatible with the survival of the cell if they were not segregated, the lysosomes illustrate the importance of physical compartmentation and division of labour. Lysosomes are discussed at length in chapter 7.

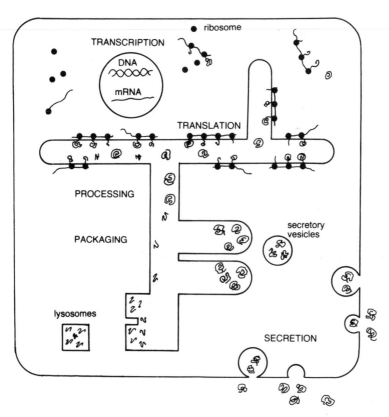

ribosome
TRANSCRIPTION
DNA
mRNA
TRANSLATION
PROCESSING
PACKAGING
secretory
vesicles
lysosomes
SECRETION

Figure 1.4 Representation of how proteins for secretion are synthesized, packaged and exocytosed in eukaryotic cells. Transcription of DNA to mRNA and translation of mRNA to protein are separated by a nucelar membrane. Proteins for plasma membrane synthesis and for secretion are threaded into the endoplasmic reticulum/Golgi system and packaged into secretory vesicles.

In both prokaryotes and eukaryotes the energy for maintenance and growth is provided mainly in the form of ATP (adenosine triphosphate). Almost all eukaryotes and many prokaryotes generate ATP by the process of oxidative phosphorylation. Chlorophyll-containing organisms are able to photophosphorylate ADP to ATP, and many microorganisms derive their ATP from fermentation or from anaerobic respiration. Although details vary, particularly among the prokaryotes, the essential molecular machinery of oxidative phosphorylation consists of electron carriers, mainly dehydrogenases and cytochromes, and ATP synthetase enzymes arranged in membranes so that the free energy change associated with electron transport from substrates to oxygen is used to phosphorylate ADP. The sites of oxidative phosphorylation and photophosphorylation in eukaryotes are the inner membrane of the mitochondrion and the inner thylakoid membranes of the chloroplast respectively. Blue-green algae have single thylakoids free in the cytoplasm. Other prokaryotes have no special organelles for oxidative ATP synthesis, and the electron carriers and ATP synthetase reside in the plasma membrane.

Most mitochondria are of the same size order as aerobic prokaryotes and there are reasons for thinking that compartmentation of oxidative phosphorylation in higher cells is the result of an endosymbiotic relationship between primitive anaerobic heterotrophs and primitive aerobic prokaryote-like cells. If so, compartmentation was probably a significant event in the progress towards the present size range of eukaryotic cells. The disadvantages that would face large cells that depend on plasma membrane-sited oxidative phosphorylation as the main ATP source can be illustrated by comparing a hypothetical cell of this type with a typical liver cell of the same size (figure 1.5). The maximum plasma membrane area available for oxidative phosphorylation in the hypothetical cell would be about 3000 μm^2 compared with a total mitochondrial inner membrane area of some 30 000 μm^2 reported for liver cells. Com-

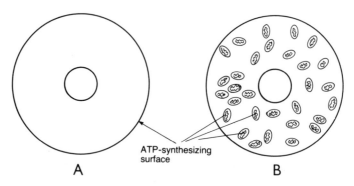

Figure 1.5 A hypothetical hepatocyte with sites for oxidative phosphorylation in the plasma membrane (A) as in prokaryotic aerobes, would have only about a tenth of the surface area provided for ATP synthesis by the total mitochondrial inner membrane area of a typical hepatocyte (B). The diffusion distances between the sites of ATP production and utilization would be much greater in A than in B.

partmentation evidently increases the potential for ATP synthesis and substantially reduces the distances over which ATP must diffuse to the energy-requiring work sites. In the case of the hypothetical cell, the average distance an ATP molecule would diffuse before it was used would be about 3·2 μm, assuming that the work sites were uniformly distributed through the cell. Using the Einstein equation for diffusion in three dimensions, and making some reasonable assumptions, this journey would take about 17 milliseconds (ms). In contrast, if the 1300 or so mitochondria in the average liver cell were randomly distributed in the cytoplasm, the average distance between the ATP synthesizing sites in the mitochondria and randomly distributed work sites in the cytoplasm would be about 0·6 μm, approximately 0·6 ms of diffusion time. In practice, mitochondria are frequently sited strategically in areas where energy is required, so the advantages of compartmentation could be even greater than these figures suggest.

Compartmentation of metabolic sequences like the tricarboxylic acid cycle and oxidative phosphorylation in mitochondria makes it possible to achieve constant and optimal concentrations of the substrates and cofactors for these sequences without imposing these conditions on the rest of the cell. This is brought about by energy-linked transport systems in the inner membrane that are able to translocate specific metabolites and nucleotides into the mitochondrial matrix, thus making them available to the enzymes connected with oxidative decarboxylation, ATP synthesis and other mitochondrial functions. However, the membrane is impermeable to various important cofactors and nucleotides that influence the rates of anabolic and catabolic reactions, e.g. NAD^+, NADH, $NADP^+$, NADPH, acetyl coenzyme A and coenzyme A. Consequently the pools of these molecules in the mitochondrial matrix are physically segregated from those in the cytosol, and can be independently controlled.

Many of the principles discussed above that show compartmentation as a necessity for large genetically sophisticated and metabolically complicated cells are underlined in the organization of chloroplasts, the chlorophyll-containing organelles which convert light energy into the chemical energy of ATP and reduce carbon dioxide and synthesize carbohydrates. The structure and biochemistry of chloroplasts is discussed in chapter 3. They are organelles enclosed by two membranes with an additional inner membrane system of flattened membrane sacs called *thylakoids* which are stacked together in colonade fashion. The thylakoid membranes contain the photosynthetic pigments and electron carriers that are involved in ATP synthesis and the generation of reduced $NADP^+$ for carbohydrate synthesis; the enzymes for carbohydrate synthesis are mostly found in the fluid stroma that surrounds the thylakoids. The chloroplast can thus be regarded as containing three compartments—a compartment between the two membranes of the envelope, the stroma compartment, and the intrathylakoid compartment. Specific translocases in the inner-envelope membrane, and transmembrane pH differences across this and the thylakoid membrane, regulate the movement of small molecules and ions between these compartments, and between the chloroplast and the cytosol, and they in their turn control the metabolic processes. Many of the small molecules which move across these membranes are phosphorylated and reduced carbon compounds, and their movement and metabolic conversion controls the energy balance between the cytosol and the chloroplast, which in fact differs in the light and the dark. The inner chloroplast envelope membrane is impermeable to NAD^+, NADH, $NADP^+$, NADPH and nucleoside di- and tri-phosphates, so the chloroplast and cytosol pools of such compounds are completely separate. In the chloroplast, then, the photophosphorylation and photosynthetic processes are neatly compartmentalized away from the mass of other cell reactions. The inner envelope membrane keeps the important intermediates of synthesis at optimal levels and prevents them diffusing into the cytosol. The thylakoid membrane arrangement provides an impressive area for photophosphorylation systems and, like the mitochondria, the organelles have some facility for being orientated in strategic intracellular positions that maximize their bioenergetic functions.

In this chapter, some aspects of compartmentation have been introduced with reference to the main organelles found in higher cells. Compartmentation and division of labour does not begin or end with organelles; it is shown at the levels of organs and tissues, whole organisms and even populations. Moving down the scale, the organelles themselves contain a host of poorly defined compartments—inter- and intra-membrane spaces, multiprotein complexes, fixed charge environments, and so on. Even the active centres of enzymes may be regarded as compartments specific for particular substrates. Compartmentation and division of labour are therefore fundamental themes in biology and sociology. However, division of labour can operate meaningfully and efficiently only if there is communication and integration between compartments, whether these are cabinet ministers or electron carriers. Although the following chapters concentrate, of necessity, on specific organelles, the fact that they are part of a greater and more complex whole should be kept firmly in mind.

CHAPTER TWO

THE NUCLEUS

2.1 Introduction

Nuclei were first described by Brown in 1833 and were quickly recognized as a constant feature of animal and plant cells. Typically they are spherical or ovoid bodies, but other shapes are not uncommon; for example, polymorphonuclear leucocytes have a lobulose nucleus of several interconnected parts, ciliate protozoans are characterized by a kidney-shaped macronucleus, and there are various shapes among the lower eukaryotes. Some nuclei, like those of the silk glands of the silk-worm, have finger-like extensions that greatly increase their surface area. In many cases, neither the biological significance of nuclear shape nor the factors causing it have been defined, although these are interesting cell biological problems. Nuclei vary in size from about 3 μm to 25 μm in diameter, depending on cell type, and contain chromosome diploid numbers that range from 6 (Indian muntjac) to 92 (*Anotomys leander,* a rodent) for mammalian species. Some polyploid species of plants and invertebrates have several hundred chromosomes. A direct relationship between chromosome ploidy and nuclear size is found in some species, e.g. the sea urchin, but not in others. In fact, there are many examples that show that nuclear size is not obligatorily linked with chromosome number or DNA content: the nuclei in different tissues of the same mammal vary considerably in size, despite having the same chromosome number. One or more nucleoli are present in the nuclei of most active tissues. These are conspicuous granular bodies without membrane, typically associated with specific chromosomes. They are the sites of ribosome assembly and are dynamic structures that change in size, disaggregating during mitosis and reforming in the daughter cells.

The nuclear envelope (perinuclear cisterna) is made of two membranes, each 5–10 nm thick, separated by a space that is up to about 50 nm wide and apparently continuous with the cisterna of the endoplasmic reticulum. The outer membrane frequently has attached ribosomes on its cytoplasmic side, whereas the inner membrane encloses the nucleoplasm and is often

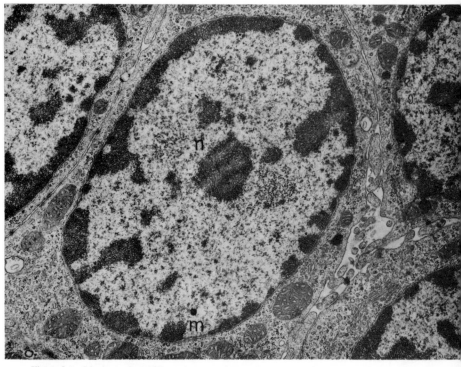

Figure 2.1 Nucleus of a cell from the zona glomerulosa of the rat adrenal body, showing the double nuclear membrane (m) and the nucleolus (n). (Dr. Gillian Bullock) (× 15 000).

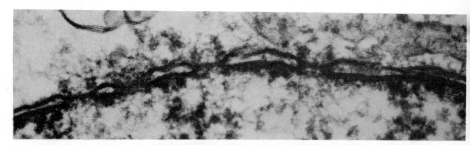

Figure 2.2 Cross-section through the nuclear membrane of a human melanoma cell in culture showing the penetration of the double membrane by pore complexes. (Dr. G. G. Maul) (× 100 000).

intimately associated with chromosome material. The inter-membrane cisterna is an intriguing compartment that has not been investigated to any extent. The nuclear envelope is traversed by pores of 65–75 nm diameter which, however, have a complex ultrastructure (section 2.9) that reduces considerably the effective diameter for transport.

The evidence for nucleo-cytoplasmic communication as a factor in cell maintenance and development pre-dates the rediscovery of Mendel's "genes" and their physical basis in the chromosomes. In the late nineteenth century, Verworm, Balbiani and others showed that, following microsurgery, nucleated halves of various protozoans survived and grew, whereas the anucleated halves degenerated and died. Later, in the 1930s it was shown that insertion of nuclei into anucleated amoebae restored pseudopodial activity, feeding behaviour and growth. It was also shown that the nucleus was essential for the growth and regeneration of the morphologically complicated ciliate *Stentor,* following removal of parts. In a classic series of experiments on the unicellular alga *Acetabularia,* Hammerling demonstrated, by means of interspecific nuclear transplants, that morphological features, notably the shape of the cap, were determined by the nucleus (figure 2.3). He also showed that, even after removal of the nucleus, the cell was able to continue morphogenesis for a time, and proposed that the cytoplasm contained a store of morphogenetic material that had been produced by the nucleus. The results of Hammerling's experiments carried out between 1934 and 1954 are, of course, now explicable in terms of the central dogma of protein synthesis developed by

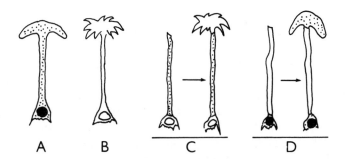

Figure 2.3 Grafting experiments on *Acetabularia.*
A: *Acetabularia mediterranea.*
B: *Acetabularia crenulata.*
C and D: grafting experiments that show that the regenerated cap is characteristic of the
nucleus and not the algal body.

the biochemical geneticists of the last decade or two, that is, that phenotypic characteristics reflect the type and amounts of proteins produced, that the genetic codes for these proteins reside in the DNA of the chromosomes, and that the molecular transcripts of the codes, in the form of messenger RNA, are synthesized in the nucleus and translated in the cytoplasm.

THE INTERPHASE NUCLEUS

The cell cycle is described in terms of four relatively short mitotic phases—prophase, metaphase, anaphase and telophase—followed by a longer interphase during which the daughter cells grow, normally doubling their mass prior to the next division. The interphase nucleus is sometimes called a resting nucleus, in contrast to the nucleus in division, but this is a misnomer since it is a hive of biosynthetic activity. These activities include the transcription of the genes into mRNA, rRNA and tRNA, the assembly and export of ribosomes, and the replication and repair of DNA.

2.2 Chromatin

Interphase chromosomes cannot be seen as discrete entities and are in the form of a fine network of chromatin fibres that makes the genes more accessible for replication and transcription than in the tightly supercoiled mitotic chromosomes. In addition to DNA, chromatin contains a small family of basic histone proteins H1, H2A, H2B, H3 and H4 (see Elgin and Weintraub (1975) for chemical details) and a large number of associated poorly-defined non-histone proteins. The amino-acid sequences of the histones and the DNA:histone ratios are very similar for chromatins from widely different sources. The chromatin fibre is composed of nucleosomes—quasi-spherical units consisting of some 200 base pairs of DNA, and two molecules each of H2A, H2B, H3 and H4—linked together by inter-nucleosomal DNA (15–100 base pairs) rather like beads on a string (figure 2.4). The DNA between the nucleosomes is associated with histone H1 molecules. The evidence for the chromatin model is as follows: mild treatment of chromatin by nucleases that cleave DNA releases spheres comparable in size (c. 8 nm diameter) to spheres observed by electron microscopy (E.M.) in negatively stained chromatin. The molecular weight of the isolated spheres is approximately the same as that calculated for the

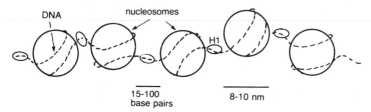

Figure 2.4 Schematic diagram of a chromatin fibre as nucleosomal beads on a DNA string. Histone H1 is shown associated with the linker DNA. Recent neutron scattering studies suggest that the nucleosomes may be discs about 11 nm in diameter and 6 nm high rather than regular spheres.

observed spheres *in situ* (160 000–180 000). The isolated spheres have DNA:histone ratios that fit the model, and addition of histones H2A, H2B, H3 and H4 to naked DNA from various sources produces chromatin with the same X-ray repeat data and E.M. characteristics as eukaryotic chromatin. This and other evidence reviewed by Kornberg (1977) put the nucleosome basis of chromatin almost beyond doubt.

The finding that certain nucleases (e.g. micrococcal DNase) digested the link DNA, but were relatively ineffective against the nucleosomal DNA, was taken at first to be consistent with earlier ideas that the histones were bound on the outside of DNA coils. Later studies including controlled digestion by proteases and nucleases and analysis of the resulting fragments showed the converse; it is now thought that the DNA is wound round the histones, possibly in a left-handed toroidal superhelix as proposed by Weintraub *et al.* (1976). Campoux (1978) has pointed out that any model of the nucleosome must accommodate a seven-fold compacting of the DNA segment and a net change in the linkage between the strands of the DNA duplex amounting to an unwinding of the helix by $1\frac{1}{4}$ turns per 200 base pairs of linear DNA. Theoretically there are many ways of doing this by combinations of superhelices and kinks. Although the pattern of the DNA binding is still obscure, progress has been made on the structure of the histone core of the nucleosome. A recent model based on studies of the dissocation and association of histone complexes is illustrated in figure 2.5. This postulates that the nucleosome core is composed of an $[\text{H3-H4}]_2$ tetramer and two H2A-H2B dimers bound together by hydrogen bonds donated by lysine or tyrosine amino acids. This histone octomer would bind and compact the nucleosomal DNA, probably by ionic interactions between residues on the histones and negative charges on the DNA.

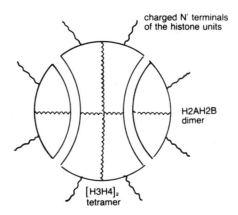

Figure 2.5 A model of the structure of the nucleosome.
It proposes that the nucleosome is made of one [H3-H4]$_2$ tetramer and two H2A-H2B dimers. The tetramer and dimers are stable because their individual subunits are held together by hydrophobic bonds (shown as wavy lines). However, the forces that form the nucleosome octameric core are weak, involving a limited number of hydrogen bonds. This means that the shape can be easily changed by ligand binding and local ionic and pH changes, and fits the concept of the nucleosome as a dynamic structure. The "tails" are the charged N terminals of the histone subunits projecting out of the hydrophobic core. They are most likely the binding sites for DNA.

All chromatins investigated to date contain nucleosomes, with the probable exception of the genes for ribosomal RNA (rRNA), but there are slight organism and even tissue-specific differences, nucleosome DNA apparently varying from 162 base pairs (rabbit cortical neurones) to 242 (sea urchin sperm). Therefore the nucleosome seems to be a universal device for compacting the long DNA molecules of eukaryotic cells. A useful account of the application of physical techniques to nucleosome structure, including nuclear magnetic resonance and neutron scattering, is given by Lilley and Pardon (1979).

2.3 Gene transcription

The copying or transcription of genes to give RNA has long been recognized as a nuclear function. RNA synthesis can be detected in most cells throughout interphase, specific RNA species being synthesized at specific times according to the gene transcription programme for the cell type. The problem of transcription raises two general questions: how genes are selected for transcription, and how transcription occurs. The main features of transcription in prokaryotes are now understood. Basically, a linear transcript of the base sequence of the gene is made by classical base pairing and polymerization of ribonucleotides, catalyzed by RNA polymerase. Only one strand of the DNA duplex is copied, and the process is facilitated and regulated by initiator and termination base sequences and protein factors for activation, suppression and completion of transcription. Eukaryotic transcription differs in detail, not least because nucleosome-organized DNA is different from prokaryotic DNA, which is not generally associated with histone. Numerous experiments have shown that non-specific transcription of chromatin *in vitro* is stimulated by removal of histones or addition of non-histone proteins. These experiments, together with kinetic data, suggest that histones block access of RNA polymerase to the DNA template, and that one or more non-histone proteins make the DNA accessible. It is not difficult to envisage how an RNA polymerase binding site might be conformationally restricted in a nucleosome, and how the experimental removal of histones or the binding of a non-histone protein could open up the site. In fact, there is evidence that the nucleosomes in transcriptionally active DNA segments are in a different conformation from nucleosomes in non-active regions. One of the attractive features of the nucleosome model outlined earlier (figure 2.5) is that the hydrogen bonding which is thought to bind the $[H3-H4]_2$ tetramer and the H2A-H2B dimers is both relatively weak and highly specific, as hydrogen bonding is markedly vectorial and, in this case, involves only a limited number of amino acids. Any modification of the hydrogen bonding capacity by, for example, ligand binding, or acetylation, or methylation of the lysine residues would change the association of the tetramer and dimers, and conceivably influence the accessibility of the DNA to RNA polymerase or other proteins that operate in transcription.

The foregoing considerations implicate the nucleosomes as dynamic rather that static structures, that play a major controlling role in transcription but leave untouched the question of how genes are switched on and off. Transcription of specific genes can be induced in cells under at least three different situations. The first is the normal programmed transcription that occurs in the course of the cell cycle—since no endogenous intracellular effectors have yet been isolated, this is a difficult system to investigate. The second case is the induction of transcription by nutrient substrates; for example, transcription of the gene for β galactosidase in yeast can be induced by addition of lactose to the medium, and various genes in the mammalian liver can be switched on by increasing levels of amino acids in the diet. This experimental system is similar to those used to demonstrate that transcription of structural genes for metabolic sequences in prokaryotes was regulated by the interaction of end products or inducers with repressor proteins specific for the polycistron involved. However, the elucidation of these "operon" systems in prokaryotes was made possible mainly by the availability of mutant strains, a technology not yet available for eukaryotic systems. A third means of inducing transcription in eukaryotes is by administration of hormones, notably steroid hormones; for example, androgen stimulates the synthesis of mRNA for prostatic fluid proteins. Oestradiol quickly induces transcription of the ovalbumen gene in chickens, and this has provided a useful system for studying gene activation. The hormone is bound by a protein receptor (MW 80 000) in the cytoplasm, and this complex enters the nucleus and binds to chromatin. It has been reported that this is followed by acetylation of histone H4 in association with transcription. Yamamoto and Alberts (1976) have proposed that the receptor protein complex binds to the chromatin at sites specified by particular base sequences, and that binding elicits conformational changes in the contiguous chromatin that makes H4 accessible for acetylation by a nuclear histone acetylase. As a result, the histone-DNA association in the nucleosomes would change to facilitate transcription. The model (figure 2.6) implies several binding sites for the receptor complex spaced over the average transcription unit of 6000–8000 bases. Experimental evidence for the model is weak, and it is included mainly to illustrate current approaches.

In prokaryotes, genes for enzymes involved in the same metabolic pathway are often clustered together, and their transcription is co-induced or co-repressed, e.g. the ten genes for proteins involved in histidine synthesis in *E. coli*. In contrast, in eukaryotes the genes for enzymes in the same metabolic pathway are typically scattered throughout the chromosomes, indicating that individual control rather than group control is the rule. Various hypotheses have been advanced that the programmed transcription that appears to be necessary for balanced growth is brought about by a "knock on" effect, for example, that transcription of a structural gene is linked to transcription of an adjacent regulator gene, the transcript or translated protein of which locates and specifically activates the next gene in the programme, possibly by binding to an initiator sequence. There are logistical problems in all hypotheses involving proteins as specific regulators, because of the large number that would be required, and the inefficiency of having regulator RNA transcripts passing out of the nucleus for translation to regulator proteins which then enter the nucleus and, as Campbell (1978) aptly puts it, search along the lengthy featureless

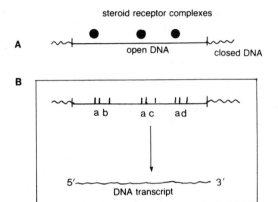

Figure 2.6 A model for gene activation.
Steroid receptor complexes bind to specific sites within the gene, thereby opening it up for
transcription as described in the text (A).
B illustrates an idea to explain how a steroid hormone complex sometimes activates different
genes at different times during development. According to this model occupation of the
specific steroid receptor complex sites *a* would be insufficient for transcription initiation unless
sites *b*, *c* and *d* were also occupied by other gene activators, the presence of which would vary
during development. The model uses combinatorial principles to generate a high degree of
regulatory specificity with a relatively small number of gene activators.

rope of chromosomal DNA for their specific binding sites. Most eukaryotic
genes are present as single copies and, if chromatin is in the form of a
random jumble, finding them would be akin to looking for a needle in a
haystack with a magnet. There is therefore increasing interest in models
that demand fewer regulator molecules and reduce the recognition
problems. In this respect, the tertiary structure of interphase chromatin is
interesting. The fact that the chromatin is in the form of large supercoiled
loops, suggests that it could be an allosteric supermolecule that can respond
to ligand binding by shape changes that could open or close down specific
blocks of structural genes that are not necessarily co-linear or even close
together. The open regions would be accessible to regulator molecules,
histone acetylases, polymerases, and other macromolecules involved in
transcription. At the time of writing there is no logically-consistent
aesthetically-satisfying model relating transcriptional control to gross
changes in chromatin shape. However, there is a correlation between
genetically inert regions of chromatin and regions of supercoiled highly-

condensed chromatin that indicates that tight supercoiling can prevent transcription; for example, the centromeric chromatin is condensed during interphase, and no transcription products of its highly reiterated short DNA sequence have been detected. Similarly the highly condensed second X chromosome (Barr body) of human female somatic cells is not transcribed. Polytene chromosomes are atypical in many ways, but the programmed puffing patterns they display make the same point—that condensed chromatin must be opened up before transcription is possible.

Before leaving the subject of transcription it should be noted that some genes, notably those for histones, ribosomal rRNA and transfer tRNA, are present in multiple copies, several hundred strong in the case of RNA. This group is sometimes called *middle repetitive sequences* because the numbers are small compared with highly repetitive short sequences that occupy a substantial part of the genome of most eukaryotes. (In the guinea pig, the sequence CCCTCA is repeated some 5 million times.) The base sequences of highly repeated segments are generally too short to function as normal genes and, until recently, they were widely held to be evolutionary deadwood. This may be the case for some repetitive sequences in some cell types but, since they occur between structural genes, and sometimes within them, and a substantial proportion is transcribed, some may well play a part in transcriptional control.

There are at least three classes of eukaryotic RNA polymerases. Evidence reviewed by Chambon (1975) indicates that class A enzymes catalyze rRNA synthesis in the nucleolus, class B enzymes are found in the nucleoplasm and the chromatin and catalyze the synthesis of the large heterogeneous precursors of mRNA, and class C polymerases are nucleoplasm-based and produce the low molecular RNA, i.e. tRNA and 5S rRNA. The evidence is compelling, but not conclusive.

2.4 Post-transcriptional modification

Many genes, including those for rRNA, ovalbumin and globins, are interrupted by sequences of DNA of unknown function ranging from several hundred to several thousand bases long. This has been shown by annealing cloned genes with their respective RNA species under conditions where stable DNA-RNA duplexes form by complementary base pairing. Using E.M. (and autoradiography) the DNA segments which cannot hybridize with the RNA can be located. These segments do not have corresponding bases on the RNA, and are consequently not encoded in the protein that results from the translation of the RNA. The question of whether the intervening sequences of the gene were looped out during the initial transcription process, or whether the entire length was transcribed and the sequences nicked out of this primary transcript to give mRNA, was partially answered in 1978 by Tilghman *et al.* Mouse β-globin gene was annealed with its 10S mRNA and also with a 15S RNA thought to be a likely precursor of the mRNA. The 10S RNA-DNA hybrid clearly showed a non-hydridized DNA loop corresponding to the non-informational piece between the two sections of the globin gene. In contrast, the entire DNA fragment hybridized with 15S RNA, showing that both structural and intervening parts of the gene were contained in this RNA, and confirming its status as a primary transcript (figure 2.7). Primary transcripts are apparently nicked, and the discontinuous segments of genetic information ligated into functional RNA in the nucleus. This implies the existence of nucleases and ligases hitherto unsuspected, possibly with auxiliary proteins to recognize the segment junctions. However, there is evidence from studies on the ovalbumin gene in Chambon's laboratory and elsewhere that the

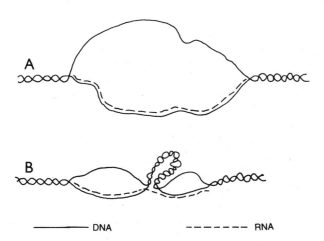

Figure 2.7 Diagram to show the principle of the experiment of Tilghman *et al.* (1978). On annealing the 15S β-globin mRNA precursor with cloned DNA containing the gene, a continuous loop is formed without evidence of intervening DNA segments (A). RNA = dashed line.
In contrast annealing the 10S mRNA with the DNA shows a looped-out section of DNA (B). The looped-out DNA represents the piece that does not carry the code for globin.

sequences at the junction have properties that simplify the process. This gene is at least 7700 base pairs long and is composed of a leader region and seven "exons" (ovalbumin coding regions) separated by "introns" (intervening sequences). The primary RNA transcript is compatible with the transcription of the entire length and, during its subsequent modification to mRNA (1872 nucleotides long), the intron transcripts appear to be removed stepwise. The intron-exon junctions of the gene have been sequenced, and it has been found that the introns start with the dinucleotides 5'-GT and finish with AG-3'. Therefore it would seem that, for ovalbumin at least, nicking of the primary RNA transcript takes place at the sequence UC/CU (the transcription sequence corresponding to AGGT of DNA), although other recognition factors may also be involved. This helps towards an understanding of the local features that define splicing but, as has recently been pointed out, the way in which the necessary long-range interactions between the ends of exons destined for ligation takes place is still obscure. An understanding of this may emerge through studies of ribonucleoprotein particles (sometimes termed ribonucleosomes) in the nucleus. These particles, which are not the easiest material to work on, are apparently made of histone-like proteins in a regular subunit organization with RNA folded around or through them. They may contain processing enzymes and proteins that influence RNA conformation to facilitate site-specific events like splicing. However, other suggestions have been put forward, including the possibilities that they are intermediates in ribosome synthesis or breakdown, or have a role in the transfer of mRNA through the nuclear membrane to the cytosol.

Most functional mRNA has a polyadenylate segment of 50–200 bases at its 3' end and a 7 methyl guanosine-pyrophosphate cap at its 5' end. The precise functions of the modifications are unknown, but they appear to stabilize the molecules *in vitro*, protect against exonuclease digestion, play a role in ribosome binding, and may have other roles. These modifications are done in the nucleus before the primary transcript is cleaved and ligated. Histone mRNA is an exception.

Modification of tRNA precursors includes the removal of 20 or so nucleotides, splicing together of the component pieces, methylation of selected bases, and other transformations to give the high and varied incidence of unusual bases characteristic of tRNA species. The tRNA precursors and their modifying enzymes have been found in both nucleus and cytoplasm.

The processing in the nucleus of rRNA precursors will be discussed in the next section in the context of ribosome assembly.

2.5 Ribosome formation

Ribosomes are the nucleoprotein particles on which translation of mRNA takes place in the cytosol and in the bioenergetic organelles. Cytosol ribosomes are rather larger than those of the chloroplasts and mitochondria, and are assembled in the nucleus. The functional cytosol ribosome is composed of two subunits: a 40S subunit containing about 30 different proteins and a molecule of 18S RNA, and a 60S subunit of some 40 proteins and a molecule each of 28S, 5·8S and 5S RNA. The ribosomal proteins are synthesized in the cytoplasm and imported into the nucleus.

There is a single primary transcript for 18S, 5·8S and 28S RNA. This is 45S RNA in mammalian cells (smaller in other eukaryotic cells), and it is synthesized and processed in the nucleolar regions of the nucleus. Nucleoli are associated with and organized by the chromosome regions that contain the genes for rRNA. Typically the rRNA genes are present as clusters of multiple copies, and their dispersion on the chromosomes determines the number and form of the nucleoli. Mutants of *Xenopus laevis* unable to synthesize rRNA have no nucleoli. During maturation of the oocytes of some species there is an enormous increase in the number of copies of the rRNA genes and, significantly, this is accompanied by an increase in the number of nucleoli. The 45S primary transcript has the general form

5′ spacer — 18S — spacer — 5·8S — spacer — 28S — 3′

After methylation of a proportion of the ribose and base residues of the rRNA segments, nuclease action cleaves out the separate segments proceeding from the 3′ to the 5′ end. The rate of processing of 45S RNA may be an important regulator of ribosome formation.

The 18S and 28S RNA form the cores of the 40S and 60S subunits respectively with proteins organized round them. The proteins, which vary in MW between about 8000 and 55 000, are present in single copies and can all be labelled or chemically modified *in situ*. This suggests that the ribosome subunit surfaces are mosaics in which at least part of each protein is present. However, ligand binding could itself alter the molecular architecture and expose hidden groups, so this conclusion is debateable. The 5S RNA of the 60S subunit is transcribed from its own gene, of which there are, if anything, even more copies than for 45S RNA. All 5S RNAs studied have a sequence complementary to the pGpCpUpA sequence in the methionine tRNA which initiates translation. On this ground it has been suggested that 5S RNA may be near the translation promoter site of the 60S subunit and involved in binding tRNA Meth. The ground is shaky.

Little is known about the processes of ribosome construction in the

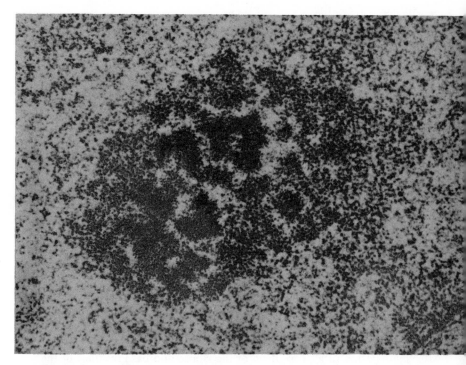

Figure 2.8 A nucleolus in a uterine glandular epithelial cell from an ovaridectomized rat treated with oestrogen for ten hours. Fibrillar and granular regions are clearly distinguishable. (Dr. D. N. Luck and Dr. J. J. Rome) ($\times$ 45 000).

nucleolus. Some nucleolar areas are fibrillar, while others have a more granular appearance (figure 2.8), the former being associated with transcription activity and the latter with ribosome maturation. E.M. studies show that these regions are not rigidly defined, and that the proportions change with the physiological state of the cell. It seems probable that, as the core RNA species become available, the ribosomal proteins are incorporated around them in a self-assembling co-operative manner. The kinetics of subunit release into the cytoplasm vary with cell type and conditions. In some cases, ribosomes are produced throughout interphase, but in others (such as the HeLa cell) transcription dominates during early interphase and maturation in late interphase.

2.6 Nuclear protein synthesis

There have been reports of protein synthesis in isolated nuclei, and it has been claimed that there is a unique class of nuclear polyribosomes. However, it is difficult to isolate nuclei free of contaminating polyribosomes, and there are many RNA-protein complexes in nuclei that might superficially give the impression of unique ribosomes, e.g. ribosomes in the process of assembly and mRNA-protein packets. No definitive biochemical evidence exists for specific nuclear polyribosomes. If protein synthesis is a general property of nuclei, its contribution to total protein synthesis must be extremely low in most cases. It could, of course, be of considerable physiological importance. Interested readers should consult the debate on this issue between Goidl and Allen (1978).

2.7 DNA replication

Studies on plant and animal cells, protozoans and eukaryotic microorganisms growing synchronously in culture, have shown that DNA doubling occurs at a specific time during interphase characteristic of the cell type. The S phases, as they are termed, of various cells relative to the preceding interphase period (G1) and the remainder of interphase (G2) are shown in table 2.1. The events that initiate DNA synthesis are not yet defined, but it is thought that a protein synthesized during G1 acts as a de-repressor, allowing DNA replication to proceed. Pulse labelling techniques have shown that each cell type has a characteristic sequence of DNA replication, some chromosomes starting long before others. Further, chromosomes do not show simple sequential replication from one end to the other, but follow ordered and complex patterns, often with numerous replication sites operating on the same chromosome at the same time. The

Table 2.1 DNA replication in the cell cycle. G1 = period between mitosis and DNA synthesis. S = DNA synthesis. G2 = period between DNA synthesis and mitosis. In most cells mitosis takes 3–15% of the G1 + S + G2 time.

	G1	S	G2	
	hours			
Tetrahymena pyriformis	0·48	0·99	0·70	Cameron & Stone, 1964
Blepharisma americanum	17	7	2	Minutoli & Hirshfield, 1968
Schizosaccharomyces pombe	—	0·17	2·5	Bostock et al., 1966
Vicia faba	4·9	7·5	4·9	Evans & Scott, 1964
Tradescantia paludosa	1·0	10·5	2·5	Wimber & Quastler, 1963
Zea mays root centre	151	9	11	Clowes, 1965
Mouse L fibroblasts	8	6	5	Killander & Zetterberg, 1965
Chinese hamster cells	2·7	5·8	2·1	Hsu et al., 1962
Newt regenerating lens	40	19	2	Zalik & Yamada, 1967.

high DNA content of eukaryotic cells makes multi-origin replication essential. The rates of chromatin replication *in vitro* indicate that up to 5000 replication sites must be operating at once in some cell types to account for genome doubling in the observed S period, which is normally quite short. The most widely used technique for investigating this is to add ³H-labelled thymidine to the culture and replace it by cold thymidine after specified periods. The chromosomes are then arrested at metaphase by the mitotic poison colchicine, and autoradiography is used to detect where the label was incorporated during the period of exposure. In recent years, more sophisticated methods have been employed, including separation of heavy labelled chromosomes and more precise evaluation of the lengths and locations of replicating sections by restriction endonuclease scissoring. An increasingly useful technique is experimentally to clone eukaryotic DNA segments in prokaryotes and other eukaryotes and follow their replication kinetics. These approaches have led to the idea of replication units or replicons as the basic elements of DNA duplication. Studies on a wide range of eukaryotes imply that most replicons are in the 10 μm ($3\cdot3 \times 10^4$ base pairs) to 100 μm ($3\cdot3 \times 10^5$ base pairs) range.

DNA synthesis apparently starts at an origin within the replicon and proceeds in both directions from this origin to the termini. The initiation sequences in SV40 and polyoma viral DNA have been identified as palindromic axes of two-fold rotational symmetry, that is, a piece of double-stranded DNA where the sequence on one strand read left to right is the same as the sequence on the other strand read right to left, e.g.

3′ A A T T G C A A T T 5′

5′ T T A A C G T T A A 3′

These sequences could conceivably form hairpin loop DNA conformations (Soeda *et al.* 1977) that would be prominent recognition sites. Proteins seem necessary to initiate DNA synthesis, since inhibition of protein synthesis stops initiation of DNA synthesis. Eukaryotic protein fractions stimulate DNA synthesis *in vitro,* catalyzed by DNA polymerases, and this, together with the discovery of DNA binding proteins that initiate DNA synthesis in prokaryotes (see Wickner, 1978), make it likely that specific initiator proteins will eventually be demonstrated in eukaryotes. The simplest model to explain early and late replicating DNA units would be a series of such proteins interacting with genetically determined origin sequences on the replicons, making the DNA accessible to the DNA polymerase.

A general model for eukaryotic DNA replication is as follows: the binding of the initiator protein to the initiator site makes the helix

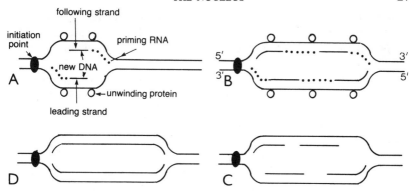

Figure 2.9 Model of DNA replication. Details in text.

accessible to DNA-unwinding proteins that open up a stretch of the duplex. A DNA-directed RNA polymerase then builds up a short complementary RNA sequence in the $5' \rightarrow 3'$ direction on each of the separated $3' \rightarrow 5'$ DNA strands. These RNA sequences serve as primers for DNA synthesis, deoxyribonucleotides being processionally added at their $3'$ ends by DNA polymerase. As figure 2.9 shows, the DNA replication forks proceed in opposite directions over the stretch opened by the unwinding proteins (polymerases always show a $5' - - \rightarrow 3'$ polarity). In the meantime, the next stretch of helix is being unwound for the process to be repeated. The short RNA primers are eventually removed by ribonuclease and replaced by DNA, and the various pieces of the new DNA strand are ligated together. This model, based mainly on the work of Kornberg and Okazaki on *E. coli,* is consistent with the finding that primary DNA products in several eukaryotic synthesizing systems are covalently linked to short RNA chains (8–10 nucleotides), and that proteins have been found in eukaryotes that perform all the functions required by the model. DNA polymerase α found in all nuclei needs a DNA template for activity and can use the $3'$ end of an oligoribonucleotide as a primer. Its activity increases in the S phase of the cell cycle, and its activity in *in vitro* systems is stimulated by addition of helix-unwinding proteins from eukaryotic nuclei. Therefore, of the three main classes of DNA polymerase, α seems the most likely candidate for a major replication role. The functions of the other polymerases, β and γ, are thought to be more related to DNA repair processes. Helix unwinding as proposed by the model would lead to torque and supercoiling, but DNA swivelases recently isolated from nuclei could solve this difficulty. These enzymes relieve torque by nicking a strand and catalyzing the transfer of the $3'$ end round the helix to close the strand on the other side.

DNA replication proceeding bidirectionally from the replicon origin in short steps is obviously a clever way to replicate both antiparallel strands by polymerases that can only operate in the 5′ − − − →3′ direction. What size are the short steps? Since DNA single-stranded fragments of 100 to 150 nucleotides have been found linked to short RNA sequences, the steps are probably upwards of 160 nucleotides. This size is consistent with evidence reviewed by Sheinin *et al.* (1978) that primary DNA intermediates of 220–280 have been isolated from eukaryotic cells during the S period. This suggests that the size of the replication steps may be related to the length of DNA associated with the basic chromatin unit, the nucleosome. (Depending on cell type the nucleosome DNA varies from 162 to 242 bases with 15 to 100 less firmly associated linker bases.) More information is required before molecular model building can be taken much further. In fact what happens to the nucleosome as the replication fork reaches it is controversial. Experiments by Weintraub indicate that the parental nucleosomes segregate on the leading strand (figure 2.9). Newly replicated DNA duplexes, depending on cell type, take up to 15 minutes to mature and acquire the nucleosome properties of stable chromatin. This underlines the point that chromatin synthesis is an integrated process that involves both DNA replication in the nucleus and histone synthesis in the cytoplasm.

2.8 DNA repair

DNA repair processes in eukaryotes were first indicated by observations that cells exposed to radiation showed survival and growth kinetics inconsistent with the calculated extent of DNA damage. It appears that an endonuclease breaks the strand near the site of damage, and an exonuclease starts at the free end and removes about a hundred nucleotides including the damaged one(s). This segment is then rebuilt by a DNA polymerase using the other strand as a template and, finally, a ligase joins the end of this repair patch to the nucleotide at which the exonuclease stopped (figure 2.10). Cell lines from people who suffer from the hereditary skin disease *Xerderma pigmentosum* have been invaluable in these studies. These cells are hypersensitive to ultra-violet (u.v.) light, in most cases because the endonuclease that makes the nick in the damaged strand to allow exonuclease to operate is not produced. Resistance to u.v. light can be restored by introducing purified endonuclease into the cells, using inactivated virus as a carrier.

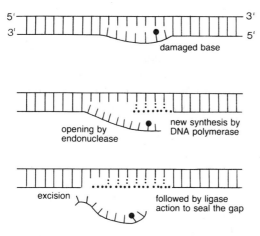

Figure 2.10   DNA repair by excision and new synthesis.

Post-replication repair sometimes occurs when damaged bases are not all removed before replication begins. In these cases, replication proceeds, leaving gaps in the daughter strand opposite the damaged bases on the parent strand that are subsequently filled. This is a little-understood area, and there are doubts about the accuracy of post-replication repairs in eukaryotes. As well as excision repair and post-replication repair, many mammalian cells can rejoin X-ray-induced single-strand breaks and, in a few reported cases, double-strand breaks. All the enzyme activities necessary for DNA repair have been found in extracts from nuclei— endonucleases, exonucleases, polymerases and polynucleotide ligases. Clearly the ability to repair DNA is important for maintaining the structure of genes. Defects in repair mechanisms may be basic to some forms of carcinogenesis, hypersensitivity to sunlight, radiotherapy and chemicals, and premature ageing.

2.9 Nucleo-cytoplasmic traffic

It will be evident from the foregoing sections that there is considerable trafficking across the nuclear envelope during interphase. Ions, nucleotides, and structural, catalytic and regulatory proteins are imported from the cytosol; mRNA, tRNA and ribosome subunits are exported to the cytosol. Although many instances of asymmetric activities of ions, non-electrolytes and enzymes have been reported, there is little evidence that the nuclear envelope has any regulative power over compartmentation of low-MW material; for example, the high intranuclear concentrations of K^+ and Na^+ are explicable in terms of the cation binding affinity of chromatin. Similarly, many enzymes found almost exclusively in the nucleus—polymerases, histone kinases, nucleases, RNA methylases and so on—are associated with the chromatin by ionic bonds or even more intimately. Also, much of the evidence for higher concentrations of amino acids and sucrose in nucleoplasm than in cytosol, fades when corrections are made for cytoplasmic space occupied by organelles. In fact, Siebert (1978) has recently reviewed evidence that nucleotides and many soluble enzymes including glycolytic enzymes have identical osmotic activities in both compartments and that the nuclear envelope *in situ* is extremely permeable (10^2 and 10^8 times more permeable than plasma membrane to glucose-6-phosphate and sucrose respectively).

In contrast to this picture for relatively low-MW material, experiments involving microinjection of tracers of known size and charge into the cytoplasm of large cells like *Amoeba* and oocytes have shown that particles of diameter greater than 10 nm are unable to enter the nucleus. Further, the nuclear envelope has a sieving effect. Paine et al. (1975) calculated that the access route was 9 nm in diameter and showed that a decrease in the diameter of tritiated dextrans from 7·1 nm to 2·4 nm increased their entry rate into oocyte nuclei over two thousand fold. Although a 9-nm-wide throughway would probably cope with most particles known to enter the nucleus, the large RNA polymerases (M.W. c. 500 000) would need to be asymmetric with a high axial ratio or enter in subunits and assemble inside. The problem is still more formidable for the 60S ribosomal subunits and high-MW mRNAs which pass from the nucleus into the cytoplasm, as the former is a fairly symmetrical unit of some 16 nm diameter and the latter can have an MW of up to 10^6.

Studies on the permeability properties of the nuclear membrane have been paralleled by studies of the ultrastructural organization of the pore. Gall described 8 annular granules, as he termed them, lining the rims of pores; with some variations this is characteristic of all pores (figure 2.11). The granules are now recognized as fibrils which, to a greater or lesser extent, traverse the pore and extend into the cytoplasmic and nuclear phases as cylindrical cages, sometimes with the ends closed like fishnets (Maul, 1977) and often attached to chromatin (figure 2.12). The traverse fibrils are stabilized by annular rings, and it seems reasonable to regard the whole as a structural and functional unit. Although it is not yet chemically defined because of isolation difficulties, it is not nuclear membrane

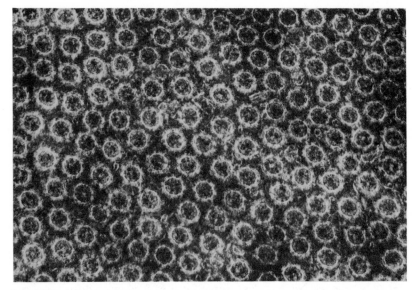

Figure 2.11 Negatively stained nuclear envelope of the oocyte of *Xenopus laevis* showing eightfold symmetry of the annular material. (Prof. Dr. W. W. Franke) (× 58 000).

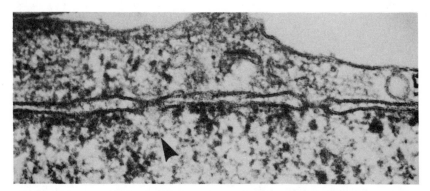

Figure 2.12a Cross-section through the nuclear envelope of a human melanoma cell in culture. In the left pore there is a ring structure (arrowed) attached to the transverse fibrils that extend into the nucleoplasm. (Dr. G. G. Maul) (× 100 000).

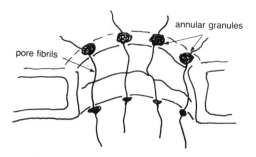

Figure 2.12b   A representation of a nuclear pore showing the granules and trans-pore fibrils.

material, as it is more resistant to detergents and sonication. In view of the organization of the pore complex and the fact that larger cellular particles cross the nuclear membrane than would be predicted from tracer studies, it seems premature to regard the pore as a static sieve tube. As we have pointed out, there is no evidence for active transport against osmotic or electric gradients. Nevertheless, some form of facilitated diffusion seems both necessary and possible. Maul (1977) has speculated that the pore fibrils might act as "tracks" or "moving ropes" to which particles are attached or indeed, that the entire complex might move through the pore. Recent evidence that ATP hydrolyzing activity is associated with pore material and that ATP is necessary for the export of mRNA from the nucleus is not inconsistent with the notion of mechanical work at the pore level driven by ATP-linked conformational change, but many other hypotheses are possible. Kinetic and biochemical data are required.

   There is an element of non-randomness in nuclear pore distribution that may be influenced by chromatin distribution, although there is no firm evidence of this. Certainly the pore population can change within hours, extended fibrillar cylinders are sometimes seen to be attached to chromatin during interphase, and pore complex structures are often associated with chromosomes during mitosis. This raises the question of whether pores might provide channels between the cytosol and specific chromatin regions. In fact, pore complexes pose many questions. Are there several populations of pores with different translocator specificities? Do ribosome subunits have pore recognition proteins? Is it easier to get out of the nucleus than into the nucleus? Are the polyadenylate tail and the protein complexing of mRNA devices to facilitate membrane translocation, as some think, or translational controls?

THE DIVIDING NUCLEUS

2.10 Chromosome segregation

During cell division the duplicated chromosomes become apparent as highly condensed bivalents or chromatids joined together at a point, the centromere. In eukaryotes, the separation of these chromatids during division is normally brought about by a protein cytoskeleton built during prophase and early metaphase. Basically, this mitotic apparatus consists of two mitotic centres and a polar spindle. Higher plants are exceptions in that they have no mitotic centres. There are various types of mitotic centres, particularly in fungi, but the standard structure in animal cells and many of the lower eukaryotes is the centriole. The centriole (figure 2.13) is a cylindrical organelle of 0·12 to 0·15 μm diameter and up to 2 μm long. Its wall is composed mainly of 9 symmetrically arranged fibres (c. 25 nm diameter) each of which is made up of 3 subfibrils. No electron-dense material has been found in the centre of centrioles, although this does not mean that they are empty. Structurally, centrioles are homologous with the kinetosomes (basal bodies) at the base of cilia and flagella. There is some

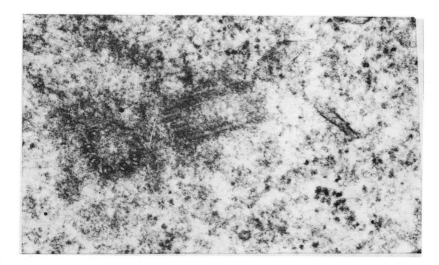

Figure 2.13   An electron micrograph of a pair of centrioles, one of which is cut in transverse section and shows internal structure. (Dr. M. S. C. Birbeck) ($\times$ 40 000).

E.M. evidence that the 9 circumferential fibres of cilia and flagella are direct extensions of the fibres of the kinetosome, although modified in some respects, e.g. they have only two subfibrils per fibre. Cilia or flagella are never found without corresponding kinetosomes. Although the role of the kinetosome is not yet understood, its association with cilia growth is in some ways analogous to the association of centrioles with the fibres of the mitotic spindle in so far as both organelles are involved in the organization of structures that are polymers of the same protein, tubulin. In some cells, aberrant kinetosomes have been observed to engage in spindle formation. Not surprisingly it has been speculated that the centriole may be an evolutionary development from the kinetoplast.

Early light microscopy indicated that mitosis started with the division of what appeared to be a single centriole into two daughter centrioles. E.M. studies have since shown that this is not division but rather the separation of two closely associated centrioles—a centriole pair or diplosome. In fact, the centriole replicates very early in interphase, sometimes before the end of telophase, and so the centriole pair is present for most of the cell cycle. The new centriole is first seen as a short "procentriole" of the same diameter and structure as the old one, but orientated at right angles to it. It is not a typical transverse fission phenomenon, and it is sometimes argued that the mature centriole induces the *de novo* assembly of the daughter centriole. The latter idea leans heavily on evidence that kinetosomes have some capacity for self-assembly, not least the fact that they appear suddenly in *Naegleria* as it changes from its amoeboid form to a typical ciliate. At present categorical statements on centriole replication, their genetic continuity and whether they contain DNA are unwise; the evidence is not firm enough. To some extent this is due to the problems of isolating these sparse and tiny organelles in pure form.

In early prophase, the centriole pair separate and migrate towards opposite sides of the nucleus, this being associated with the appearance of connecting fibres. These are composed of microtubules (MTs) and associated proteins. MTs are of variable length and about 20 nm in diameter; their walls are made of helically arranged globular proteins called tubulin (MW *c.* 120 000). The polymerization of tubulin into the connecting MTs is probably the force that propels the centrioles to their respective poles. As they move, each centriole acquires an ever-increasing aster or centrosphere—satellite material that contains granules, membrane vesicles and disorganized MTs. Eventually the centrioles stop at each side of the nucleus and MTs, growing out of the centrospheres, penetrate the nuclear envelope. In most eukaryotes the envelope disintegrates, and tubulin subunits from the cytoplasm enter the nuclear area. The MTs growing from each pole make contact and interdigitate to form the mitotic spindle. Continuous pole-to-pole MTs are formed in many fungi and protozoans, but in most eukaryotes the two sets of MTs overlap but do not extend over the whole length. Each chromatid becomes attached to the nearest pole by MTs between the centrosphere and the kinetochore, a specialized protein region firmly embedded in the chromatin fibres of the chromatid centromere. E.M. studies have shown that kinetochores are growth points for MTs and that isolated chromosomes can polymerize purified tubulin *in vitro*. Therefore it seems that both mitotic centres and kinetochores are involved in assembling the thirty to fifty MTs that eventually link each chromatid with its polar region. Little is known about the physical and chemical factors that influence MT orientation. Tubulin must be activated before it can be polymerized, a process that involves phosphorylation by the nucleoside triphosphate GTP, probably by protein kinase action. One idea is that mitotic centres release a protein kinase or kinase activating factor that diffuses out, polymerizing and orientating the tubulin as it goes; the resulting MTs would tend to radiate with the advancing diffusion front, and the spindle would form as the fronts meet. This model obviously requires that the main tubulin pool is in the central region of the cell, otherwise there would be radially symmetrical MT growth round each mitotic centre. E.M. observations implicate the centrosphere rather than the centriole as the main organizer of MT production. This is supported by evidence from *in vitro* experiments on pericentriolar material isolated from hamster ovary cells (Gould & Borisy, 1977) and the fact that, although angiosperms lack centrioles, they form spindles from ill-defined centriosphere-like centres. Conceivably centrospheres might produce a diffusible activating factor, but it is also possible that their main role in spindle formation is to provide the initial binding and nucleation sites for tubulin, and that orientation of the ensuing MTs into a spindle has nothing to do with hypothetical diffusing activators. MTs have a propensity for growing relatively straight and high tubulin concentrations in the central region of the cell could well be the dictating influence for spindle form. There is good E.M. evidence that the main spindle fibres serve as a reference for the orientation of MTs growing from the kinetochores to their respective poles. Sister kinetochores are on opposite faces of the centromere that joins the chromatids, and this may help their association with opposite poles.

The bivalents that are randomly distributed at the beginning of prophase are brought into dynamic equilibrium at the cell equator prior to their

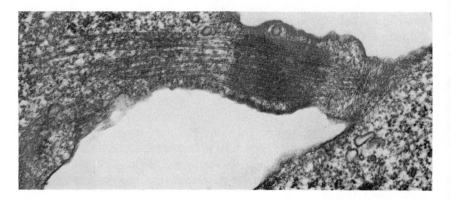

Figure 2.14   A thin section through a primary muscle cell culture (from new born rat thigh) where myoblasts are in process of dividing and fusing into multinucleate myotubes. The section shows chromosomal material aggregated on a mitotic spindle. An aster is shown on the right of the chromosomes near the junction of the dividing myoblast with the adjacent cell material. The spindle fibres to the left of the chromosomes are less restricted physically and extend further. (Dr. D. Lawson) (× 40 000).

segregation at anaphase. The process involves much pulling and jerking of the bivalents, making and breaking of MTs and a progressive increase in the number of MTs. At metaphase (figure 2.14) when the maximum MT concentration is achieved, the combined microtubular length can be as great as 4–5 cm for large cells like sea urchin eggs.

During anaphase the chromatids segregate. In most eukaryotes this is effected by a shortening of the MTs attaching the chromatids to the poles, and a further separation of the poles themselves. One hypothesis is that subunits are progressively removed from the pole-chromatid MTs and are built into the pole-pole MTs, thereby shortening one set and elongating the other. Alov and Lyubskii (1977) have speculated that the kinetochore might control the depolymerization of the pole-chromatid MTs and that the resulting increase in free tubulin would lead to polymerization of pole-pole material to maintain the dynamic polymer/monomer equilibrium. The model is compatible with many observations but lacks definitive proof. There is growing evidence that most eukaryotes except fungi and ciliates have relatively few continuous pole-pole MTs, and that the MTs from each pole stop short, giving an overlap but not continuity. In this context

McIntosh has proposed that the two sets of spindle filaments may move over one another by a type of ratchet mechanism or, more specifically, by the making and breaking of intertubular cross bridges using ATP as the energy source (figure 2.15). The idea is rather similar to the sliding-filament model of muscle contraction but less developed. This model has some attractive features. It fits the analysis of MT numbers in some types of cells obtained by serial sectioning at different division stages, for example, the data of Tippit *et al.* (1978) on the diatom *Fragilaria*. Lateral fibres arising from MTs seemingly in a cross-bridge position have been observed in the spindle of some cells. Griffith and Pollard (1978) have shown that MTs, associated spindle protein and actin interact *in vitro* in a manner that suggests that they could interact in an ATP-dependent force generating system *in vivo*. Although this work is in its preliminary stages, it is significant because actin and myosin have both been demonstrated in spindles, as has ATP hydrolyzing activity. A further sign that the sliding filament model is a serious contender is recent and substantial evidence that flagellar motion is due to mechanochemical actions of dynein cross-bridges between the MTs. In conclusion, no single model fits the range of mitotic patterns that exist in eukaryotes. Kinetic cross-bridge models do not easily explain pole-chromatid MT shortening, and it may be that different cell types exhibit different degrees of dependence on mechanochemical action and assembly-disassembly principles. Alternatively, new unifying hypotheses could emerge. In the end, models stand or fall on the basis of experiments.

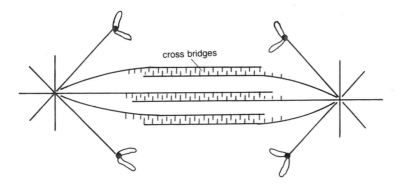

Figure 2.15   Sliding-filament model for chromosome separation.

2.11 Nuclear membrane breakdown

In some of the lower eukaryotes, cell division occurs without nuclear membrane breakdown (discussed in chapter 8). However, "open division" with nuclear membrane dissolution is the preferred mechanism for organisms with larger genomes. The main reason for this may be that it is easier to handle larger numbers of chromosomes by direct attachment to mitotic spindles rather than through a membrane intermediary, but more subtle factors may be involved. The breakdown of the nuclear membrane results in a mixing of the nuclear and cytoplasmic constituents—an efficient way of bringing tubulin synthesized in the cytoplasm to the spindle construction zone and releasing ribosomes in the opposite direction. Also the structure of the nucleus in "open dividers" may be such that it is easier for it to be disintegrated and rebuilt than to undergo binary cleavage. Relatively little is known about the factors that control nuclear membrane breakdown, but it is clear that cytoplasmic factors are involved. Cytoplasm from maturing oocytes can induce chromosome condensation and nuclear membrane breakdown when injected into immature oocytes. The appearance of this maturation promoter factor in maturing eggs is correlated with the time of nuclear membrane breakdown; it is synthesized even when the nucleus is experimentally removed, and its appearance is prevented by inhibition of protein synthesis (Wasserman and Smith, 1978).

One of the first signs of nuclear breakdown is the disappearance of nuclear membrane pores. Maul (1977) reports that in less than an hour (prophase to prometaphase) almost the entire 4000 pores disappear from the nuclear membranes of cultured mammalian cells. It is not clear whether the pore structure is dissolved or conserved; as mentioned before, pore complexes have been found on chromatin during mitosis. Later the entire membrane disintegrates, so that during anaphase it cannot be detected. During telophase the membrane begins to reappear, initially without pores. However, once the elementary structure of the membrane is complete there is very rapid pore formation, suggesting assembly of presynthesized units rather than new synthesis and incorporation.

CHAPTER THREE

PLASTIDS AND CHLOROPLASTS

3.1 Structure of plastids in higher plants

The possession of at least one form of plastid within its cytoplasm has been suggested as the feature which most clearly distinguishes a eukaryotic plant cell from an animal cell. Certainly no living cell of a higher plant has so far been described which completely lacks plastids. Plastids are often more or less spherical or disc-shaped (1 μm to 1 mm in diameter), but may be elongated or lobed and frequently show amoeboid characteristics, particularly in developing tissue. They can be distinguished from other cellular organelles by their double bounding membranes, the possession of plastoglobuli (spherical lipid droplets) and by the presence of an internal membrane fretwork of several discrete internal vesicles. Every tissue of a plant has more than one plastid type and during the life cycle of a plant as many as 10 to 15 different plastids may function in its cells. The life cycle of a radish seedling is shown in figure 3.1 to illustrate the variety of plastids in its tissues. The chloroplasts—plastids whose internal membranes contain chlorophyll—are functionally and structurally the most complex plastids and, because of their unique role in mediating photosynthesis, have been the most thoroughly investigated. A myriad of other forms of plastid have also been recognized by morphologists and given names largely based on their structural features or on the pigments they contain. Many were first described by the eighteenth and nineteenth-century microscopists. Their appearance in thin sections is illustrated in figure 3.2.

Proplastids are morphologically the simplest form of plastid and are found in the young cells of roots, stems and leaves, particularly in the dividing cells of the meristem. Cells in the shoot meristem have approximately 7 to 20 proplastids, but up to 40 may be present in root cap cells. Proplastids (figure 3.2) are between 1 and 3 μm in length and may be spherical or ellipsoidal and are often also amoeboid. Their few internal membranes are rounded or spherical vesicles. Crystals and starch grains are often present in the proplastid matrix, particularly in root proplastids. Despite their small size and simple structure, proplastids can be readily

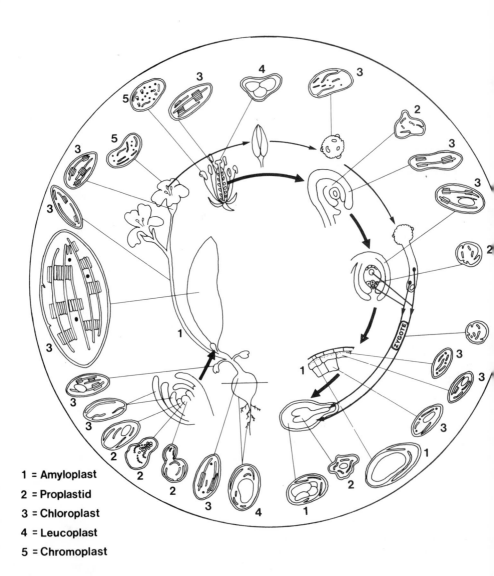

Figure 3.1 A diagram showing the type and distribution of plastids found in the organs of a higher plant during successive stages in the plant life cycle from seed development to fertilization. (The diagram is modified and redrawn from Tageeva *et al.* (1971) in *Photosynthesis and Solar Energy Utilization*, ed. O.V. Zalensky (Hayka, Academy of Sciences, U.S.S.R.)

1 = Amyloplast
2 = Proplastid
3 = Chloroplast
4 = Leucoplast
5 = Chromoplast

distinguished from mitochondria because of the presence of plastoglobuli and the close alignment of the two envelope membranes. The proplastids in the young cells of the leaf typically differentiate into chloroplasts, and in corn leaves the differentiation is complete in less than three hours. Plastid differentiation is accompanied by rapid chlorophyll synthesis and a six-fold increase in plastid volume, a massive proliferation of internal membranes, and rapid biosynthesis of macromolecules such as proteins and nucleic acids. Methods have been developed for the isolation of proplastids but very little is known of their biochemical function. The name "proplastid" almost certainly describes a heterogeneous category of organelle, and there is no evidence that all proplastids in tissues other than leaves have the potential to develop into chloroplasts.

Etioplasts are the plastids found in the cells of the leaves of plants which have grown for several days in complete darkness (figure 3.2). They develop from proplastids and at "maturity" are elongated in profile, between 4 and 8 μm in length and 2 to 4 μm in width. They are characterized by the presence of a central prolamellar body, a paracrystalline structure of cross-connecting membrane tubules which often almost fill the plastid. These membranes have no chlorophyll but contain small amounts of the non-reduced unphytylated precursor protochlorophyllide. Etioplasts contain 70S ribosomes, DNA fibrils, the coupling factor for photophosphorylation and Fraction 1 protein. In these characteristics they resemble chloroplasts.

Etiochloroplasts are plastids developing in previously etiolated leaves which have now been exposed to light. They are functionally and structurally intermediate between etioplasts and fully photosynthetic chloroplasts, and differ markedly from developing proplastids because of their greater structural complexity.

Chromoplasts are plastids found in petals, fruits and some roots and, because of the carotenoid pigments they contain, give a red, orange or yellow colouration to these tissues. During their development from chloroplasts, chlorophyll is destroyed and large amounts of carotenoids synthesized. Examples of tissues which contain numerous chromoplasts are the flowers of buttercups (*Ranunculus*) and nasturtium (*Tropaeolum*) and the roots of carrots. The chromoplasts in red pepper fruits (*Capsicum*) contain over 30 different carotenoids. Chromoplasts are much larger than chloroplasts and may be lens-shaped, spindle-shaped, rounded or elongated, and are often amoeboid. They are filled with plastoglobuli, some up to 100 μm in diameter, and also often contain crystals or numerous closely packed 1 μm fibres.

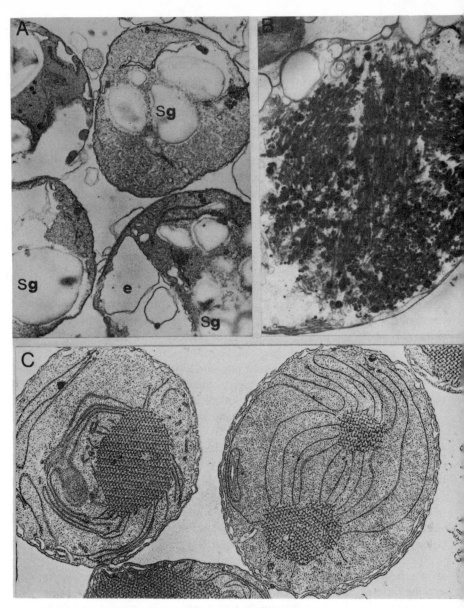

Figure 3.2 A. Electron micrograph of proplastids isolated from root cells of *Vicia faba* L. starch grain (sg), swollen envelope (e) (from Thomson *et al.* (1972) *Plant Physiology*, **49,** 270–272. (× 22 500).
B. Electron micrograph of a red pepper chromoplast showing internal fibres (photograph Dr. B. E. Juniper) (× 25 000).
C. Electron micrograph of isolated etioplasts from *Zea mays* (corn) (R. M. Leech) (× 15 000).

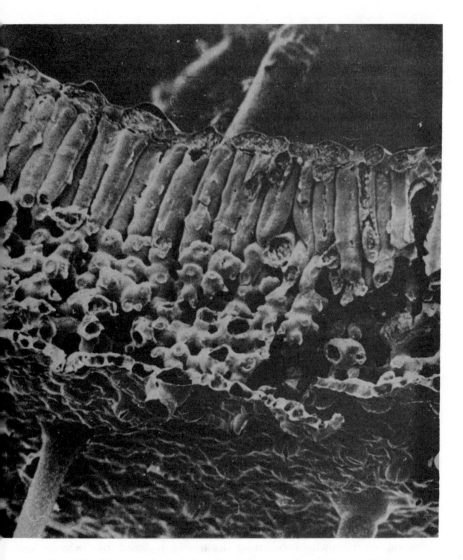

Figure 3.3 Steroscan electron microscope picture of a potato leaf (× 1050).

Plastids with special contents. Several types of plastids have been given names which describe their contents.

Amyloplasts are mature plastids in which most of the matrix is filled with starch. Amyloplasts may contain one (as in the potato tuber) or many (as in pea roots) starch grains.

Elaioplasts are plastids containing large amounts of lipids in osmiophilic plastoglobuli which often almost completely fill the plastid stroma. They are particularly common in cacti and in the epidermal cells of many lilies and orchids.

Leucoplasts are colourless plastids which do not contain starch or other deposits. They are found typically in leaf epidermal cells and the colourless storage tissue of many fruits.

Plastids respond very rapidly to changes in their cellular environment. Mineral deficiency or herbicide toxicity, for example, are often detectable by morphological changes in the structure of the leaf chloroplasts before other cellular changes can be observed. It has also been recognized for over 100 years that many of the morphological forms of plastids are interconvertible, for example chloroplasts can change to chromoplasts and vice versa and amyloplasts in some leaves are a constant intermediate stage in the development of a proplastid into a chloroplast. The plastids in the cells of the rind of navel oranges are an excellent example of chloroplast/chromoplast interconversion. As the fruit ripens, the reticulated fretwork of the chloroplast internal membrane system becomes reorganized into a complex of fibres and lipid droplets. Concomitantly chlorophyll is broken down and the rapid synthesis of carotenoids, particularly β-carotene occurs and the green chloroplast changes into an orange chromoplast. Occasionally in commercial orchards the reverse process occurs and an orange-coloured orange changes into a green one as the rind chromoplasts revert again to chloroplasts. This change can represent financial disaster to the grower so it is fortunate that treatment with ethylene usually transforms the chloroplasts again to chromoplasts.

Chloroplasts

Chloroplasts are the most intensively studied and well-known of the plastids, and the rest of this chapter will be devoted to a discussion of their cell biology. As mediators of photosynthesis, chloroplasts are responsible by the reduction of carbon dioxide for the production of all the carbon chains which provide food for the world and also the continuing supply of oxygen to the atmosphere. The frequently posed question: "why is the world green?" would be more accurately replaced by "why is the chloroplast green?". Chloroplasts are green because they contain two types of chlorophyll molecules, chlorophylls a and b which constitute the light-harvesting and energy-transduction systems of photosynthesis. The chlorophylls are Mg tetrapyrrole porphyrins, and the resonating structure at the centre of the molecule results in its ability to absorb red and blue light, and to reflect green light. However, chloroplasts are not merely machines for the photoreduction of carbon dioxide—they have a host of other metabolic functions in the green cell. Chloroplasts are able to synthesize many small molecules, notably triose and pentose sugars, amino acids and fatty acids, and to mediate the terminal stages in the biosynthesis of chlorophyll, nucleic acids, starch, complex lipids and polypeptides, and in the assembly of proteins from subunits. During photosynthesis, the chloroplasts provide a constant supply of triose phosphate to the cell cytoplasm at rates as high as 1500 μmoles C/mg chl/hr. Triose phosphate is a key intermediate in many of the metabolic pathways located in the cell cytoplasm, and its supply is constantly renewed during photosynthesis. Small molecules can also enter the chloroplast from the cytoplasm and be further elaborated endogenously. One example is the ketoacid α-oxoglutarate which the chloroplast is unable to synthesize, but which can enter the chloroplast from the cytoplasm and be aminated to glutamate. There are many examples of the collaborative function of chloroplasts with other cellular organelles and with the cytosol in biosynthesis, and a picture is emerging of cooperation of metabolic function in which the whole cell is involved. Indeed there is abundant evidence that the development of the chloroplast phenotype itself is most certainly the result of interaction between the chloroplast and nuclear genomes. Chloroplasts possess their own DNA and RNA which are chemically distinct from the cytoplasmic components, and chloroplasts are also able to synthesize nucleic acids and proteins. It is now clear that the plastid genotype is contributed to by both plastid and nuclear DNA, and that several of the plastid proteins encoded for in the nuclear DNA are synthesized on cytoplasmic ribosomes. Before describing the details of these interactions, the structure of the chloroplast will be considered.

3.2 Structure of chloroplasts in higher plants

The leaves of higher plants generally possess at least two types of chloroplast-containing cells. In many temperate plants, these can be distinguished as an upper palisade mesophyll, and a lower spongy mesophyll layer. The shapes and distributions of these cells are shown in the stereoscan micrograph of the potato leaf in figure 3.3. In living cells the shape of the chloroplast is constantly changing, and the surface and subsurface regions of the chloroplast are continually mobile. Often protuberances are temporarily produced, and changes in shape and size can be induced by osmotic changes, and also under the influence of light or as a result of cytoplasmic streaming. The bean *Phaseolus vulgaris* possesses chloroplasts which are typical of the sun leaves of higher plants: there are about 45 chloroplasts per palisade mesophyll and 32 per spongy mesophyll cell. This means that the whole leaf contains about $8 \cdot 3 \times 10^8$ chloroplasts and there are about $2 \cdot 3 \times 10^7$ per square centimetre of leaf tissue. The number of chloroplasts per cell varies from one species to another, and there are wide differences, for example, maize *Zea mays* has about 30 chloroplasts per mature leaf cell, while some varieties of wheat have 200. Chloroplasts are typically discoid or lens-shaped, with a diameter of *c.* 5 μm and a length of about 10 μm.

The dynamic aspects of chloroplast structure are lost during the fixation procedures of electron microscopy, but electron micrographs reveal much more of the fine structural detail of the chloroplasts. In thin sections showing profiles, four constant features can always be recognized in higher plant chloroplasts. The chloroplast is bounded by an *envelope* of two membranes, the inner of which encloses the *stroma* or hydrophilic proteinaceous matrix of the chloroplast. Within the matrix is embedded the fenestrated continuous membrane system containing the chlorophyll, the *thylakoid fretwork* and, between the membranes but not within the

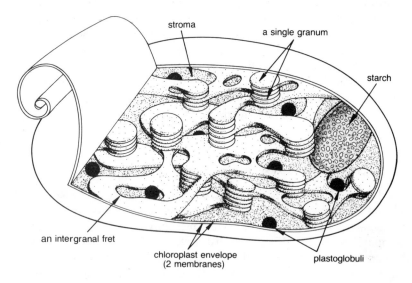

Figure 3.4   Diagram of a model of a young chloroplast showing the main structural features.

thylakoids, are the osmiophilic droplets known as *plastoglobuli*. These features are illustrated in figure 3.4.

3.2.1 *The Chloroplast Envelope*

The outer chloroplast envelope membrane is permeable to solutes apparently non-selectively. In contrast the inner envelope membrane shows very specific permeability properties and contains at least four specific anion translocation systems.

1. A phosphate translocator—facilitating an exchange of inorganic phosphate with 3-phosphoglycerate or dihydroxyacetone phosphate.
2. A dicarboxylate translocator—which facilities the counter exchange of dicarboxylic acids, particularly oxaloacetate and malate.
3. An ATP translocator which operates ten times more slowly than (1) and (2) but may be responsible for the entry of ATP in the dark.
4. A bicarbonate translocator.

It is now possible to isolate and characterize the envelope membrane from chloroplasts. So far no suspensions containing either the outer or inner envelope membranes alone have been obtained, so all our information refers to a mixed population of envelope membranes. The first enzyme to be shown to be specifically associated with chloroplast envelope membranes was the enzyme responsible for the incorporation of UDP-galactose into galactolipids (Douce *et al.*, 1974). There is also good evidence that the nitrate reductase of the leaf cell is associated with the chloroplast envelope membranes (Rathnam, 1978).

In the past it has sometimes been suggested that the inner thylakoid fretwork may develop from the inner envelope membrane. This suggestion was based on a rather limited number of observations of a close association between the two membranes, particularly in young plastids; the idea does not have wide support, particularly in view of the very different lipid and protein amino-acid compositions of the two types of membrane.

3.2.2 *The chloroplast stroma*

In profiles of chloroplasts, the stroma appears granular, the granules being heterogeneous in size and staining properties. Some of the granules are recognizable as Fraction 1 protein or ribosomes, sometimes in polysomal configurations. Other particles are presumably the "soluble" enzymes of the chloroplast stroma, for example, those of the photosynthetic carbon reduction cycle. The complex designated the ATP-synthetase can be shown as a protuberance of the membranes at the top and bottom of the grana, and on the intergranal connections, in whole mounts of negatively stained membranes viewed in the electron microscope.

In most chloroplasts, Fraction 1 protein accounts for at least 50% of the soluble stroma protein. It is the most abundant protein in the world and, on purification, is shown to consist of two types of subunit and to possess the enzyme activity of the photosynthetic carboxylating enzyme. It was originally known as ribulose diphosphate carboxylase or carboxydismutase, but recently has been renamed ribulose bisphosphate carboxylase-oxygenase (RubisP carboxylase) to reflect its ability to react with both carbon dioxide and with oxygen. The protein, measured either by its enzyme activity or electrophoretic mobility, is absent from the mesophyll cell chloroplasts of C-4 plants, although it is present in abundance in the bundle-sheath cells of these plants. It has been shown that the mesophyll chloroplast DNA contains the gene for the large subunit but that there is no large subunit mRNA in these cells. The control of the large subunit gene in C-4 mesophyll cells is thus at the level of its mRNA synthesis.

It is important to consider a few of the characteristics of Fraction 1 protein since it plays such an important role in both photosynthesis and photorespiration. The protein has a molecular weight of approximately 5.25×10^5 daltons and consists of 8 large (MW $c.$ $5.2–6 \times 10^4$ daltons) and 8 small (MW $1.2–1.8 \times 10^4$) subunits. A model based on X-ray diffraction data showing the most likely arrangement of the subunits is given in figure 3.5. Tryptic peptide analyses, immunological studies and amino-acid

A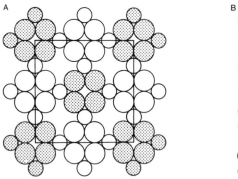

B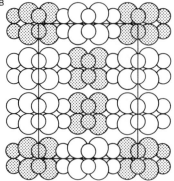

Figure 3.5   Diagram showing a model of the large and small subunits of Fraction I protein assembled in a crystal structure (from Baker *et al.* (1977) *Science* **196**, 293–295). Ribulose biphosphate carboxylase: A two-layered, square-shaped model of symmetry 422).

analyses in different species of the genus *Nicotiana* (tobacco) show that the large subunits of several species are closely related chemically, but that each large subunit shows no similarities with the small subunit of the same species. Antisera to the large but not the small subunit inhibit the enzymic activity of the complete protein, so it is possible that the active site is on the large subunit. The small subunits probably constitute the regulatory site.

The mode of synthesis of Fraction 1 protein is also particularly interesting since it requires cooperation between nuclear and chloroplast genomes. The small subunit is encoded in the nuclear DNA, and its precursor (MW 20 000) is synthesized on 80S cytoplasmic ribosomes; the large subunit is coded for by the chloroplast genome and synthesized on 70S chloroplast ribosomes. At present the mode of transport of the precursor molecules across the chloroplast envelope is unknown, but it is clear that the processing to the small subunit and the assembly of the 2 subunits occurs inside chloroplasts (see Ellis *et al.*, 1978, for a review).

The catalytic function of Fraction 1 protein will be considered later.

Chloroplast ribosomes

The biogenesis of chloroplast ribosomes is another example of cooperation between nuclear and chloroplast genomes. The genes for all the chloroplast ribosomal RNAs are found in the chloroplast DNA, and the genes for the 5S, 16S and 23S rRNA components of the chloroplast ribosome are transcribed as a single unit to give a precursor of MW 2.7×10^6 daltons which is synthesized by isolated chloroplasts. The chloroplast DNA also contains the genes for chloroplast tRNAs. Chloroplast ribosomes contain about 90 proteins and, although many of the genes for ribosomal proteins are in the nuclear genome and are inherited in a Mendelian fashion, on present evidence it seems that some of the genes at least are in the chloroplast DNA.

Chloroplast ribosomes resemble prokaryotic ribosomes in their sedimentation coefficients and RNA subunits, the sensitivity of their protein synthesizing mechanisms to the inhibitor D-threo-chloramphenicol, and their insensitivity to cyclohexamide (an inhibitor of cytoplasmic protein synthesis). Chloroplast protein synthesis is stimulated by light or ATP and, in addition to the proteins already considered, over 100 different polypeptides are synthesized by chloroplast ribosomes, the majority of which are unidentified. Those identified include the α, β and DCCD-binding components of the chloroplast ATP synthetase complex, several thylakoid proteins, and cytochrome f.

Chloroplast DNA

In profiles of chloroplasts examined in electron micrographs, a number of independent DNA-containing areas are frequently seen in the stroma of the chloroplast. The number of such areas seems to vary considerably: in some plastids there may be only one area, and in very large chloroplasts as many as 32. The amount of DNA per plastid in higher plants has been calculated to be $2–40 \times 10^{-15}$ g. Calculations of the genome size suggest that the DNA is present in about 10–30 multiple copies and that the chloroplasts can be regarded as polyploid. Chloroplast DNA is a covalently-linked naked circular molecule of molecular weight $85–95 \times 10^6$ daltons and at least 80% of the base sequences in each circle are unique. In some higher plants, each circle of chloroplast DNA contains 2 copies of an identical sequence arranged in an inverted orientation (see figure 3.6). One of the unique characteristics of chloroplast DNA is the lack of methylated bases. Examples have already been given of the complex way in which the chloroplast genetic system interacts with the nuclear/cytoplasmic one in leaf cells. DNA is replicated within the chloroplast in a semi-conservative manner. Using similar techniques to those which have been used to map mitochondrial DNA, the first stages in the mapping of chloroplast DNA are now complete. A general structure is shown in figure 3.6. Chloroplast DNA is envisaged to contain three components: component A which is

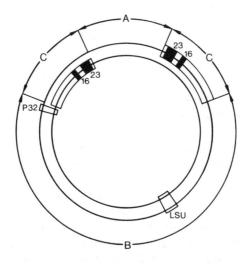

Figure 3.6 A general structure for chloroplast DNA (with permission from T. A. Dyer, from Bedbrook *et al.* (1980) Proc. 4th John Innes Symposium). P32: genes for 32 000 dalton membrane protein; LSU: gene for large subunit of ribulose bisphosphate carboxylase; 23 and 16: genes for the ribosomal 23S and 16S RNAs respectively.

very variable in length and single copy DNA; component B which is relatively constant in length and which is single copy and contains the genes for the 32 000 dalton membrane protein (P32) and the gene for the large subunit of ribulose bisphosphate caryboxylase (LS); and component C which is the inverted repeat and contains the ribosomal RNA genes. It is interesting that in *Chlamydomonas reinhardii* the direction of transcription of the ribosomal genes relative to the LS gene is in opposite orientation to that in *Zea mays* and *Spinacia oleracea.*

Temporary inclusions of the chloroplast stroma
Starch grains
In some chloroplasts, starch grains are the single largest inclusion. They appear in profile as oval structures, generally electron-translucent, but occasionally electron-opaque. In rapidly photosynthesizing chloroplasts, the presence of starch is ephemeral and, if plants are grown under low light regimes, no starch accumulates. Starch accumulation is influenced by many environmental and cellular factors, of which phosphate balance in the cells appears to be one of the more critical (see chapter 8).

Crystalline inclusions
A variety of crystals, some of which may be proteins, others of mineral origin, have frequently been reported in chloroplasts of many species. Phytoferritin, an iron-protein complex, may also be sometimes seen as a crystalline or loose cluster, but it can always be identified because the iron-containing apoprotein appears as an electron-opaque particle without osmium fixation.

3.2.3 *The chloroplast granal fretwork system*

The 3-dimensional arrangement of the internal pigmented membrane system of a higher-plant chloroplast is intricate and highly ordered (see figure 3.7). The reticulated membrane fretwork is an integrated system and can be isolated as a single unit from the chloroplast. The fretwork is usually considered to be a complex of membrane-bound vesicles, the thylakoids, which in some parts of the fretwork lie in closely adjacent parallel stacks known as *grana* which alternate with regions containing separate protrusions from the granal thylakoids. These form the intergranal connections or *frets*. Frets can be regarded as flexible channels, and one or several may connect with each granal compartment. The frets run in all directions and form a 3-dimensional mesh linking the grana together. Serial sections have shown that the frets may be arranged in a spiral fashion

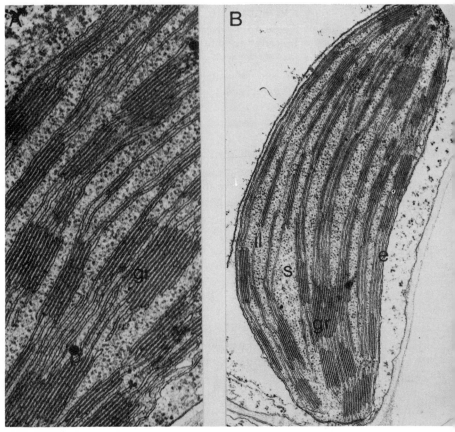

Figure 3.7 Electron micrograph sections of (A) the thylakoid fretwork (Dr. W. W. Thomson) (× 20 000) and (B) a chloroplast profile (Dr. W. W. Thomson) (× 10 000). grana (gr), plastoglobular (p), stroma (s), envelope (e).

around the granum, and it is possible that several helices may occur on the same granum. The grana themselves may be arranged in a variety of orientations to the long axis of the chloroplast. In the majority of sections of chloroplasts, grana are seen in many different views—in face view, in surface view, and in a variety of oblique section. Grana vary greatly in shape and dimensions: in some chloroplasts the few huge grana fill the whole interior of the chloroplast, whilst in others the granal fretwork is made up of many small grana of perhaps only 2 or 3 compartments. Many published electronmicrographs show an unusual situation where most of the partitions run approximately parallel with the long axis of the

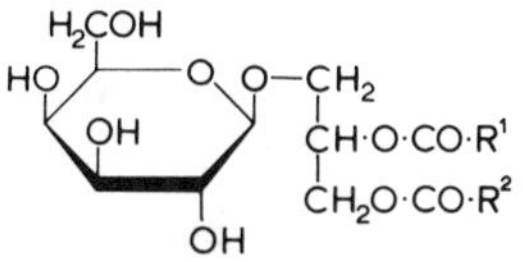

monogalactosyl diglyceride

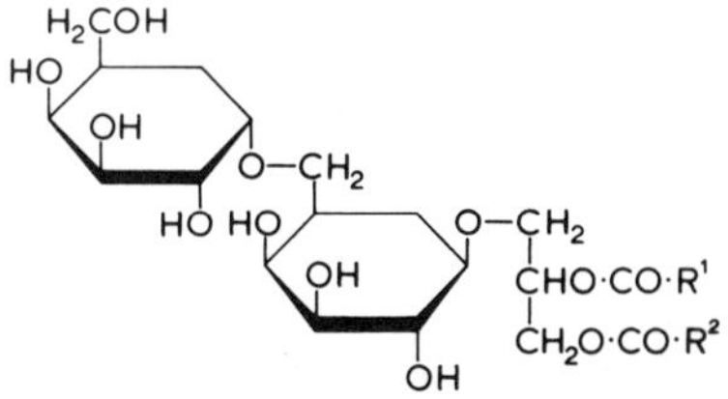

digalactosyl diglyceride

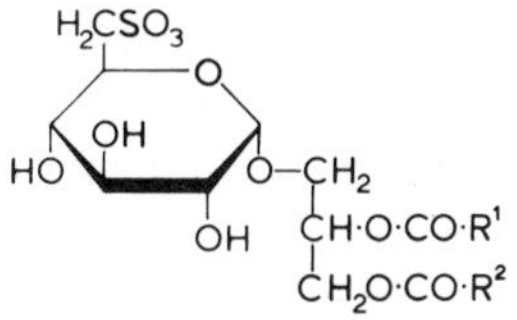

sulphoquinovosyl diglyceride

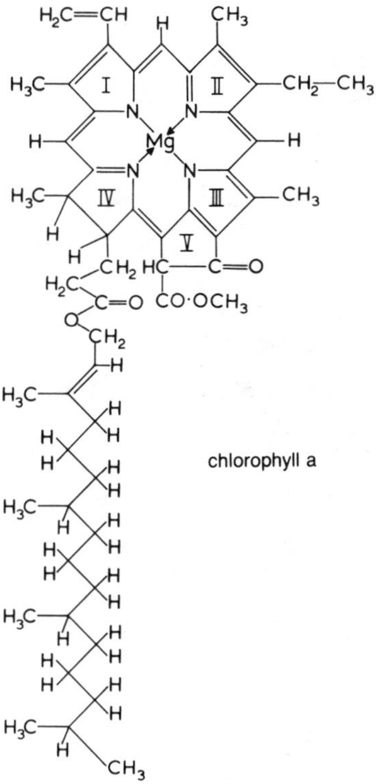

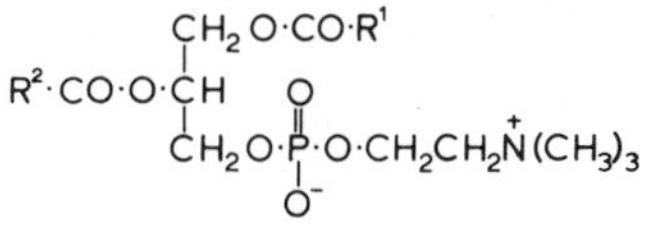

phosphatidyl choline

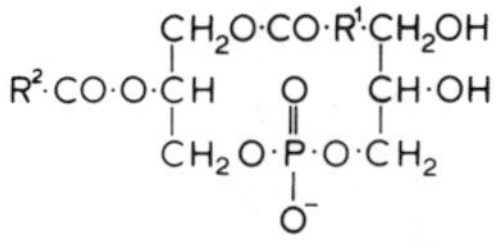

phosphatidyl glycerol

chlorophyll a

Figure 3.8 The structure of the major lipids of the chloroplast and of chlorophyll.

chloroplast. These pictures are chosen rather more for their artisitic merit than as typical of the structures they portray.

Since the chlorophyll-containing membranes of the granal fretwork are the site of the light-harvesting and energy-transduction reactions of photosynthesis, their molecular architecture has been most intensively studied. Current models visualize the membranes as fluid mosaics made up of galactolipids and phospholipids in which the particles containing proteins, chlorophyll and the components of the electron-transport pathway are embedded. The thylakoid membranes are 50% protein and 50% lipid, and the lipids are peculiar in that the majority of the molecules are galactose-containing and all have an extremely high content of polyunsaturated fatty acids (figure 3.8). The trienoic acid, α-linolenic acid, represents 90% of the lipid of pigmented chloroplast membranes. Phosphatidyl glycerol is an important lipid component of these membranes which also contain a small but constant quantity of sulphoquinovosyl diglyceride. The lipids are visualized as being arranged in a bimolecular lipid layer with their polar groups external in the hydrophilic environment of the stroma, and their hydrophobic tails in the interior of the membrane (figure 3.9). There have been no serious attempts yet to explain the exact

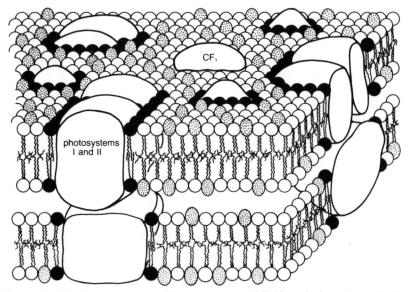

Figure 3.9   Model showing a possible arrangement of lipids and photosystems in the thylakoid membrane.   ⬡ digalactosyl diglyceride, ○ monogalactosyl diglyceride, ● phosphatidyl glycerol. CF_1 is part of the ATP synthetase complex.

arrangement of these components in the partition regions of the grana where two membranes, each a bimolecular layer, are opposed together. In freeze fracture pictures of chloroplast membranes, particles can clearly be seen, and these are generally of two size categories—one category $c.9$–11 nm in diameter, and larger particles of diameter 15–18 nm. The distribution and groupings of these particles are different in different regions of the chloroplast membrane system. In general, the intergranal connections have more of the smaller particles more widely spaced than in the granal membranes which contain about equal numbers of large and small particles. Many suggestions have been made about the composition and function of these particles, but none of the conjectures so far are supported by conclusive experimental evidence. Almost certainly they contain thylakoid proteins, and the chlorophylls and electron-transport components of the 2 photosystems. Staehelin et al. (1976) have provided an excellent review of this subject.

3.2.4 Plastoglobuli

Plastoglobuli are large lipid droplets, found singly or in groups, embedded in the stroma between the granal membranes. In sections of chloroplasts they are circular in outline and between 10 and 500 nm in diameter. Their physical integrity is retained after isolation, and they are found to contain a variety of lipidic and lipophilic compounds, particularly those also found in the chloroplast membranes. In green actively photosynthesizing leaves they do not contain chlorophyll or carotenoids but, as the plastid begins to senesce, the plastoglobuli grow larger and in chromoplasts the plastoglobuli may become a dominant feature as the carotenoid pigments accumulate within them. In developing plastids there are usually only a few small plastoglobuli. Plastoglobuli are found in all plastids and apparently represent a store of lipids surplus to current requirements.

3.3 Chloroplast development

Chloroplasts in leaf cells develop from proplastids in the meristematic cells of the leaf primordia. The ultrastructural changes which occur during chloropast development have been followed in expanding green bean leaves (Phaseolus), and also in the developing cells of monocotyledonous leaves growing from an intercalary meristem. In Phaseolus, the developing proplastids accumulate several round starch grains which they lose as they pass into an amoeboid stage, which is followed by a phase of extensive development of the perforated thylakoids and the formation of simple grana. By seven days after germination, the thylakoid system has become a continuous fretwork, and several grana are recognizable. Subsequently the plastid increases in overall size, and in the size and number of its grana. In corn *Zea mays*, where a gradient of plastid development can be followed in cells increasingly further from the leaf base, the chloroplast can develop from a proplastid in 3 hours, and the ultrastructural changes which the plastid passes through during development are generally the same as those observed in *Phaseolus*. In corn, the chloroplast increases 10 times in volume during development, the ribosome numbers per plastid increase fourfold, and the chlorophyll and membrane lipids increase tenfold. The young chloroplasts of grass leaves typically divide during development, after cell division has ceased, and the chloroplast number increases three to four fold per cell. When does a chloroplast become photosynthetically competent? The answer seems to be: as soon as chlorophyll is detectable in the developing proplastid. The light-harvesting and energy-transduction and electron-transport reactions can all be demonstrated at a stage when the internal membranes are still sparse, and the membrane associations limited to the formation of bithylakoids.

Preliminary studies have already shown that the chloroplast products of gene expression change greatly both quantitatively and qualitatively during leaf development and, in the case of the production of Fraction 1 protein, there is some evidence that these changes may be linked to changes in mRNA.

The development of the etiochloroplast

During recovery from etiolation, after illumination, chlorophyll biosynthesis and the onset of photosynthetic function take place within the etioplast. Because of the ease with which etiolated greening plants can be manipulated, they have been extensively investigated as model systems for the study of chloroplast development. It is already clear that etiochloroplast development differs in several respects from normal chloroplast development, but the characteristics of the former may indicate some of the changes which may occur in the much more complex normal plastid development. When etiolated plants are illuminated, protochlorophyllide which has accumulated in the prolamellar body is reduced to chlorophyllide, which is esterified with geranyl geraniol, which is subsequently saturated to phytol to give chlorophyll. Further synthesis of chlorophyll is stimulated, the prolamellar body loses its regular structure, additional membrane synthesis is stimulated, and finally a chloroplast exhibiting the normal structures is formed. When etiolation is reduced to a convenient minimum, and high humidity is maintained during illumination, there is no lag phase in chlorophyll synthesis in etiochloroplasts, and photochemical activities can be measured after a few minutes of illumination. Cyclic photophosphorylation can be measured more or less instantly, and the other photochemical activities associated with photosystem I appear after a few minutes, and photosystem II activity (oxygen evolution) can be measured after 30 minutes (see p. 60). Analysis of the etioplasts before illumination established that they contain all the components necessary for photosynthetic electron transport except chlorophyll; once the terminal steps of chlorophyll synthesis are completed, then photosystem I is functional. After an average of about 1 hour of illumination, photosystem I activity can be detected, and after 1–2 hours CO_2 fixation can be observed. By this stage protein and lipid biosynthesis is also considerably stimulated.

Since chlorophyll biosynthesis seems to be the key which turns on the photosynthetic function of the etiolated plastid, the details of its formation will be briefly considered.

The biogenesis of chlorophyll
Several recent changes in our understanding of chlorophyll biosynthesis have occurred. The five-carbon compound δ-aminolevulinic acid (ALA) is the first committed metabolite in the biosynthetic pathway of chlorophyll, as it is in the synthesis of other porphyrins. However, in greening leaves, the origin of ALA appears to be from the carbon skeleton of glutamate (Beale *et al.*, 1975) via α-oxoglutarate and α, δ-dioxovalerate. The postulated sequence of events is shown in figure 3.10. The ALA is then converted by the enzyme ALA dehydratase to porphobilinogen. The details of the enzymology between porphobilinogen and the magnesium vinyl pheoporphyrin A-5 (protochlorophyllide) are not yet known in detail, but a generally accepted pathway based partly on work on haem biosynthesis is shown in figure 3.10. Protochlorophyllide itself is associated with a protein known as holochrome, the apoprotein is similar to Fraction 1 protein but antigenetically distinct from it. Apparently 5 to 25 chromophores, each of MW 6×10^5 make up each unit for photoconversion. Protochloro-phyllide/holochrome can be removed from prolamellar body membranes by aqueous buffers of low ionic strength, and therefore appears to be an extrinsic protein. The photoreduction of protochlorophyllide to chlorophyllide occurs only while the pigment is associated with the protein,

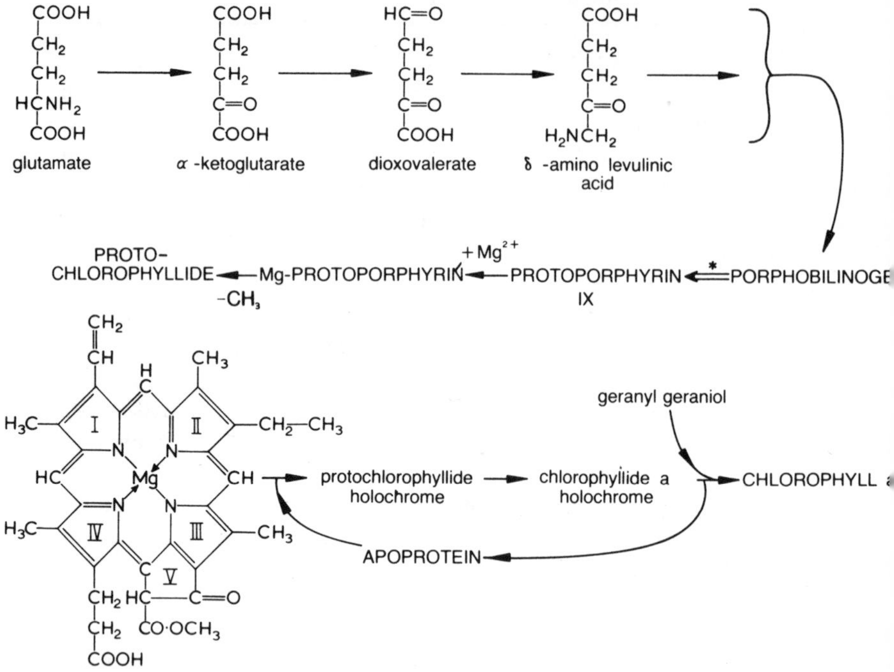

Figure 3.10 The pathway of chlorophyll biosynthesis in higher plants as in haem biosynthesis.

but has been shown to take place in intact leaves, isolated etioplasts, and also in isolated holochrome preparations.

The chlorophyllide is subsequently converted to chlorophyll a by the esterification with geranyl geraniol which is then saturated. Previously the chlorophyllide was thought to be esterified directly with phytol. During these last stages of chlorophyll formation, additional spectral shifts occur which have been interpreted as indicating conformational changes as the chlorophylls become inserted in the membrane in a variety of regular and irregular associations. In mature chloroplasts only two chemically distinct chlorophylls, chlorophyll b and chlorophyll a, are present in the lamellae, but at least ten spectrally distinct forms of chlorophyll a with different red absorption maxima can be recognized. These have been reported to have maxima between 662 nm and 700 nm, and their presence in the membranes of an etiochloroplast represents the final stage in its development to a chloroplast.

3.4 Chloroplast photosynthesis

The unique function which sets apart the chloroplast from other cellular organelles is its ability to mediate photosynthesis. For convenience, photosynthesis may be considered to consist of three separate phases:

(a) The light-harvesting and energy-transduction stage
(b) The generation of assimilatory power (ATP and NADPH)
(c) The reduction of CO_2 to triosephosphate using ATP and NADPH.

All these processes occur simultaneously during photosynthesis which may be best summarized in a single equation which describes the stoichiometry of the process:

$$2H_2O + CO_2 \xrightarrow[\text{chlorophyll}]{hv} CH_2O + O_2 + H_2O$$

Photosynthesis will occur only in the presence of light and chlorophyll to yield a carbohydrate (CH_2O) and oxygen derived by photolysis from water. None of the oxygen evolved is derived from CO_2. Many complex biophysical and biochemical events are summarized in this equation: those associated with the light-harvesting, energy transduction and photophosphorylation occur in the thylakoid membrane, and CO_2 reduction takes place in the chloroplast stroma compartment (figure 3.11). The biophysical and biochemical events of photosynthesis also occur at different rates. Each photochemical event occurs in $c.10^{-12}$ second; the light-harvesting and energy-transduction stage is completed in 10^{-9}

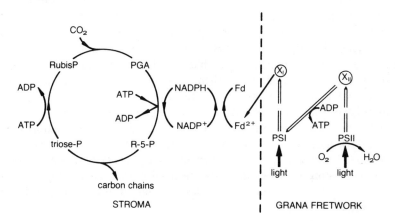

Figure 3.11 Diagram showing the inter-relationships between the three stages of photosynthesis within the chloroplast (Dr. S. M. Ridley).

second. The rate-limiting step in the electron-transfer reactions leading to the generation of NADPH takes 10^{-2} second for completion, which is also the average timing for the enzymic reactions involved in the reduction of the CO_2.

3.4.1 *Light-harvesting and energy transduction*

The absorption of light leading to the excitation of chlorophyll is the primary event of photosynthesis.* The specific arrangements of the chlorophyll molecules in the thylakoid membrane, and the nature of the association between the lipids and proteins within them, facilitate these primary events and also allow subsequent energy transduction to occur. In view of the significance of the biophysical and biochemical events which occur within them, it is hardly surprising that the details of the molecular architecture of the green membranes of chloroplasts have attracted such interest in recent years. Within these membranes, chlorophyll b and at least 10 spectrally distinct forms of chlorophyll a with absorption maxima *in vivo* between 662 nm and 700 nm are implicated in the excitation process. The forms of chlorophyll a *in vivo* apparently result from an association with different protein molecules and from specific orientations of chlorophyll molecules within the membranes.

Light conversion in photosynthesis may be regarded as a cooperative event in which many chlorophyll molecules participate together as a unit. It is generally accepted that there are two photoreactions in photosynthesis operating in series, that their mechanisms differ, and that the centres in the membranes where the photoreactions occur differ both in the assembly and the function of the pigment molecules within them. The general terms photosystem I (PSI) and photosystem II (PSII) are used to describe both the two distinct pigment assemblies and the processes of energy transduction associated with them (figure 3.12). The majority of chlorophyll b molecules are associated exclusively in photosystem II. Photosystem I absorbs light at longer wavelengths than photosystem II, and for this reason is often called the "red" system. The majority of chlorophyll molecules in each photosystem function as light-harvesting "antennae" molecules, but the heart of each photosystem is a reaction centre consisting of fewer than 10 specifically orientated chlorophyll a molecules. These chlorophyll molecules are referred to as P700 in PSI and P690 in PSII.

* There is no evidence from higher plant chloroplasts that the carotenoids of the thylakoid membrane are involved in the light-harvesting and transduction phases of photosynthesis. It seems clear, however, that they do have a protective function, preventing the photo-oxidation of chlorophyll molecules.

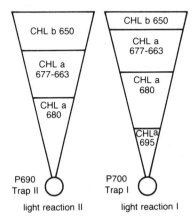

Figure 3.12 Diagram to show a possible distribution of the forms of chlorophyll between PSI and PSII (modified from R. P. F. Gregory (1977) *Biochemistry of Photosynthesis*, 2nd edition. Wiley).

The evidence points strongly to a molecular organization in which the chlorophyll molecules are specifically conjugated to different polypeptide chains, and suggests that these chlorophyll-protein complexes (or chlorophyllins) are aggregated in an array containing on average 400 chlorophyll molecules. Very recently, using improved fractionation techniques, Thornber's group in Los Angeles has provided evidence for at least three complexes in the thylakoid of green plants:

1. A chlorophyllin complex largely composed of P700-chlorophyll a protein and containing c.30% of the total chlorophyll. This complex is associated with PSI activity.
2. A complex containing 2 or more chlorophyllins but not yet specifically characterized. This complex is associated with PSII activity.
3. A light-harvesting complex primarily composed of the light-harvesting chlorophyll a/b protein (50–60% of the total chlorophyll) but with additional minor chlorophyll a and chlorophyll b-containing chlorophyllins.

The P700-chlorophyll a and light-harvesting chlorophyll a/b protein had previously been distinguished by their different apparent sizes on polyacrylamide gel electrophoresis of 90–120 kD and 22–35 kD respectively. A tentative model which best fits the present data and describes a possible molecular arrangement of these complexes within the membranes is shown in figure 3.13. The light-harvesting a/b protein and the P700-chlorophyll a protein have been well characterized chemically, and their roles in the primary photochemical events suggested. The light-harvesting a/b protein contains the bulk of the antennae chlorophyll molecules which harvest the light, while P700-chlorophyll a protein

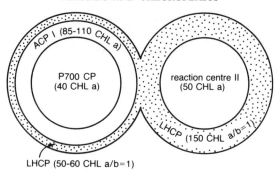

LHCP (50-60 CHL a/b=1)

Figure 3.13   A model for the organization of the chlorophyll/protein complexes in the photosynthetic unit of higher plants (from Boardman *et al.* (1978) *Current Topics in Bioenergetics* **8**, 35–109). ACP I = chlorophyll protein complex I.

resembles the heart of photosystem I, one of the two photosystems functionally recognizable. The reaction centre of photosystem II is extremely labile, and very difficult to isolate, and at present we have no information about its structure. There is also no definite knowledge of the detailed arrangement of the lipid molecules within the thylakoid membrane. Anderson (1975*a*, *b*) has made some suggestions about how the chlorophyll molecules may be orientated with respect to the proteins within the membrane. She has postulated that the bulk of the antennae chlorophyll is in the fixed boundary lipids that form a monomolecular layer attached to the hydrophobic shells of the intrinsic proteins. There is no convincing experimental evidence supporting or contradicting this interesting model, although in the bacterial photosynthetic reaction centre the chlorophyll is embedded within the protein.

Although ideas about the precise molecular architecture of the photosystems are necessarily speculative at present, conclusions about the primary photochemistry occurring within them have been derived from spectroscopic investigations of both intact plants and also isolated sub-membrane assemblies exhibiting different functional activities. Chlorophyll looks green in white light, because it absorbs light in the red and blue parts of the spectrum, and transmits and reflects in the green. Each quantum (photon) of red light* which is absorbed by a chlorophyll molecule, raises an electron within the molecule from a ground state to an

---

* Blue light contains more energy/quantum than red light and causes even greater excitation of the chlorophyll molecules, but the elevated electron falls back into the "red" orbit too quickly to allow useful chemical work. Whatever the quality of light absorbed, the electron reaches the same energy level more or less immediately after excitation, and all subsequent events are derived from this common starting point.

excited state, and all the available energy is transferred in the process. The excitation is essentially an oxidation process. Photons absorbed by any of the 400 or so antennae molecules are rapidly transferred (in about 10^{-12} second) from one chlorophyll molecule to another, and eventually trapped in the reaction centre: the whole process takes about 10^{-9} second. In detergent micelles containing chlorophyll at the same concentration as found in the thylakoid, transfer of excitation energy occurs by inductive resonance without the production of intermediate photons. This is the most likely mechanism for energy transfer *in vivo*.

In the reaction centres of PSI and PSII, the transfer of an electron to an acceptor molecule causes it to become reduced, and the chlorophyll donor molecule to become oxidized. Chlorophyll can be re-reduced and returned to the ground state if it accepts an electron from a suitable donor. If the acceptor molecule is at a lower electrochemical potential than the excited state of chlorophyll, the transfer will occur spontaneously. In fact, two sequences of electron carriers of increasingly positive oxidation-reduction potential carry electrons from the reduced acceptors of the reaction centres of photosystem I and photosystem II respectively. Although the chemical identity of the acceptor molecules is unknown, their reduction initiates the biochemical events of photosynthesis.

The operational relationship between the two photosystems and the carriers in photosynthetic electron transport is generally represented by the so-called Z-scheme originally proposed by Hill and Bendall in their classic paper of 1960. The carriers identified in the chloroplast thylakoid membrane include cytochromes f, b_{559} and b_6, plastoquinone, plastocyanin (a copper protein), ferredoxin (a non-haem iron sulphur protein), and several other iron sulphur proteins. There is a consensus about the role of the majority of these carriers in photosynthetic electron transfer, but the precise function of the b cytochromes is still open to debate.

Concomitantly with the passage of electrons via the operation of the Z-scheme, coupled synthesis of ATP occurs. This is noncyclic photophosphorylation: cyclic photophosphorylation is associated with cyclic electron flow from reduced ferredoxin back to chlorophyll a in PSI. The characteristics of ATP formation in photophosphorylation are consistent with the operation of a chemiosmotic mechanism as originally suggested by Mitchell in 1961. A more detailed discussion of Mitchell's ideas is given in the following chapter. As his theory applies to photosynthesis, the flow of electrons through the carriers of the chloroplast thylakoid membrane is believed to create an electrochemical potential

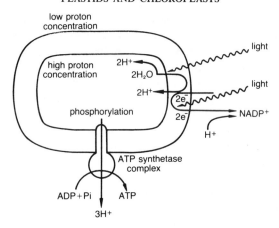

Figure 3.14   Diagram to show the relationship between the Z-scheme, the translocation of protons and photophosphorylation in the chloroplast thylakoid membrane (modified from Hinkle and McCarty (1978) *Scientific American.*

difference of protons. An asymmetric arrangement of molecules across the membrane allows this proton difference to become established (figure 3.14). Consensus of opinion based on theoretical considerations, and the results of inhibition experiments using antibodies to specific carriers, support an asymmetric arrangement of carrier molecules across the thylakoid membranes as summarized in the two-dimensional diagram in figure 3.15. and in figure 3.16. The synthesis of ATP is driven by a reverse flow of protons and involves the operation of an ATP synthetase complex (figure. 3.14). An elegant experiment by Hind and Jagendorf in 1963 first demonstrated that the establishment of this proton difference, induced by a pH change across the thylakoid membrane, could indeed result in ATP synthesis. In their experiment, transfer of isolated thylakoid membranes from a suspending medium with a pH of 6 to one with a pH of 8 resulted in ATP formation in the dark. Numerous experiments have subsequently confirmed this observation and that the intrathylakoid pH changes during illumination, and have also shown that the formation of ATP is correlated with the dissipation of this artificially produced pH change. The effect of "uncouplers" of photophosphorylation can also be explained if they are considered as proton ionophores causing a dissipation or neutralization of the potential difference. While it is generally accepted that chemiosmotic energy can be formed at some stage between the electron-transport reactions and the formation of ATP, and that this is reversible, the exact mode of conversion of hydrogen ion movement to ATP is unknown.

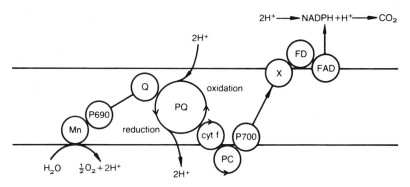

Figure 3.15 An asymmetric arrangement of electron transport carriers in the thylakoid membrane (modified from the suggestion of Trebst (1974), *Annual Review Plant Physiology* **25**, 423–458).

An ATP synthetase complex couples the diffusion of protons back through the membrane with ATP synthesis. Five subunits of this complex have been identified so far, and the superficial part of the complex (CF_1) can be visualized as an 11-nm knob protruding above the thylakoid membrane in freeze-fracture pictures. The other portion of the complex (CF_0) is embedded within the thylakoid membrane and is part of it.

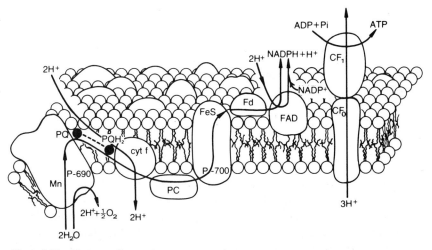

Figure 3.16 Diagram of a possible arrangement of the photosystems and electron carriers in the end granal thylakoid membrane of a higher plant chloroplast (modified from Hinkle and McCarty (1978) *Scientific American* **238**, p. 104).

Jagendorf's group have shown that when chloroplast membranes are held at pH 4 and then rapidly brought to pH 8, a hundred molecules of ATP are synthesized per CF_1 complex. The ATP synthetase complex has a reversible function and can also act as an ATPase. It is generally accepted that 3 protons are required for each ATP synthesized.

3.4.2 *The utilization of photosynthetic assimilatory power (ATP and NADPH) in carbon dioxide reduction*

All plants utilize the ATP and NADPH produced in the photochemical and electron-transfer events of photosynthesis in the reductive pentosephosphate or Benson-Calvin pathway of CO_2 reduction. Provided a supply of ATP and NADPH is available, all the reactions of carbon reduction will occur in the dark. The net product of the reduction of 3 molecules of CO_2 is one molecule of triosephosphate (and one molecule of inorganic phosphate enters into organic combination); the sequence of enzymic reactions leading to this synthesis is shown in figure 3.17. Assimilatory power is used at two points in the cycle: in the formation of the acceptor molecule ribulosebisphosphate (RubisP) from ribulosemonophosphate (RuMP), and in the conversion of the first product of carboxylation, phosphoglyceric acid (3-PGA) into triosephosphate (TP). The stoichiometry of the cycle can be appreciated by considering the fate of three molecules of CO_2 as in the diagram below:

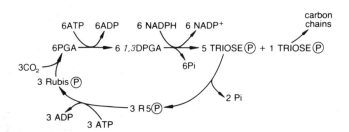

The carboxylation of three molecules of CO_2 by three molecules of the five-carbon sugar, ribulosebisphosphate, catalyzed by the enzyme ribulosebisphosphate carboxylase/oxygenase gives rise to six molecules of phosphoglyceric acid which is phosphorylated and reduced to six molecules of triosephosphate (with concomitant Pi release). One of these six molecules, the "product", probably as dihydroxyacetone phosphate (DHAP), is rapidly exported from the chloroplast, and the other five

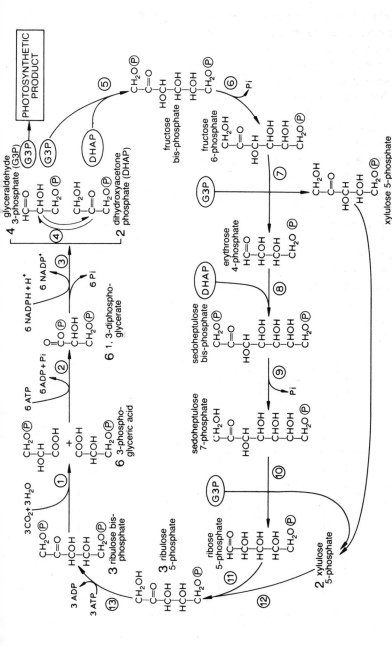

Figure 3.17 The carbon pathway of photosynthesis—the Benson-Calvin pathway. The diagram shows the fate of the six triosephosphate molecules synthesized during the photoreduction of 3 molecules of CO_2.

Enzymes involved in the pathway: 1, ribulose bisphosphate carboxylase; 2, phosphoglycerokinase; 3, triosephosphate dehydrogenase; 4, phosphotriose isomerase; 5, aldolase; 6, fructose bisphosphatase; 7, transketolase; 8, aldolase; 9, sedoheptulose bisphosphatase; 10, transketolase; 11, phosphopentose isomerase; 12, phosphopentose epimerase; 13, phosphopentokinase.

triosephosphate molecules undergo a series of sugar interconversions, collectively resulting in the regeneration of three more molecules of the acceptor ribulosebisphosphate.

The rapidly photosynthesizing chloroplast is therefore dependent on its environment for light, and for an adequate supply of CO_2 and inorganic phosphate. In its turn it provides the cell cytosol with a constant supply of small carbon molecules (Stocking and Larson, 1969), which are key metabolic intermediates, and which are also phosphorylated and reduced. Triosephosphates can move into and out of the chloroplast at rates in excess of 500 μmoles/mg chl/hr. Since the average rate of photosynthesis in a temperate plant is 100–150 μmoles CO_2/mg chl/hr it is clear that the transport process cannot be rate-limiting. It has been shown that the movement of triosephosphate and inorganic phosphate across the inner chloroplast envelope membrane is facilitated by the operation of a specific reversible translocase (Heldt and Rapley, 1970) which operates in the manner shown in figure 3.18. The respective rates of the two aldol condensations and the two transketolase reactions of the triosephosphate acceptor regeneration cycle compared with the rate of export of triosephosphate to the cytoplasm, will determine the balance between regeneration and export of carbon within the photosynthesizing cell. As previously mentioned, the light, CO_2 and inorganic phosphate

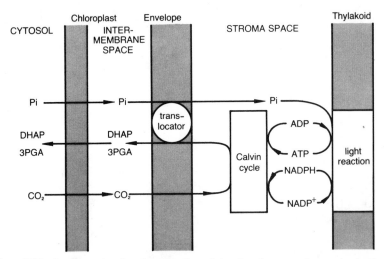

Figure 3.18 A scheme showing the operation of the phosphate translocator in the inner chloroplast envelope membrane (adapted from Heldt (1976) in *The Intact Chloroplast*, ed. J. A. Barber, Elsevier, p. 215).

concentrations will also exert an independent regulatory role on the photosynthetic process itself. Export of carbon from the cell to other tissues of the plant is necessary for growth, and another level of regulation of carbon movement in plants is in response to source/sink changes between its different tissues. Since there is no evidence that pyridine nucleotides and nucleoside mono, di, and triphosphates can enter the chloroplast at any appreciable rate, so the endogeneous and exogeneous chloroplast pools of these compounds are metabolically separated. The recognition of the significance of triosephosphate movement has been one of the most exciting advances of recent years, as it was not previously clear by what mechanism phosphorylated and reducing equivalents resulting from the light reactions of photosynthesis could be made available to the rest of the cell (figure 3.19).

Within the chloroplast itself, the regulation of the integration of the parts of the photosynthetic process, and the control of the balance between the different phases which are both temporally and spatially separated, is subject to numerous regulatory controls. An example will illustrate the complexity of some of these interactions. When leaf tissues or intact isolated chloroplasts are brightly illuminated, they do not immediately begin to fix carbon dioxide. There is an induction period before maximal rates of photosynthesis are achieved after a few minutes. During this light

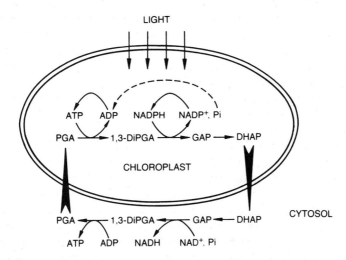

Figure 3.19 A scheme for the operation of the dihydroxyacetone phosphate/phosphoglycerate shuttle in the light (from Krause and Heber (1976) in *The Intact Chloroplast*, ed. J. A. Barber, Elsevier, p. 187). GAP = glyceraldehyde-3-phosphate (G3P).

phase, in addition to enzyme activation (e.g. of RubisP carboxylase by Mg^{++} ions), the cycle intermediates build up, and the 3-PGA → DPGA conversion plays a key role in this induction and also in the regulation of the cycle activity. Walker and Robinson (1978) have recently shown that as RubisP builds up to steady-state levels during the lag phase, the ATP/ADP ratio at first rises to 1·44 and then falls to 0·26 in the steady state, i.e. at first more ATP is formed than is utilized in the PGA → DPGA conversion. As more ribulose monophosphate becomes available, more ATP is used in RubisP formation, the ATP/ADP ratio falls and slows the conversion of PGA to DPGA. High 3-PGA concentrations are therefore needed during active photosynthesis to maintain the conversion against the unfavourable equilibrium. Rapid removal of DHAP, controlled by cytosolic inorganic phosphate levels during active photosynthesis, is therefore necessary to prevent build-up of pentosephosphate, which would in turn inhibit PGA reduction. Other examples of the regulatory interactions between chloroplast and cytosol are given in chapter 8.

3.4.3 "C-4 photosynthesis"

All green plants operate the Benson-Calvin pathway leading to the immediate end product, triosephosphate, a C-3 compound, but some plants possess an additional pathway which yields a C-4 compound as the first product. These latter plants have become known as "C-4 plants" to distinguish them from the temperate plants producing only triosephosphate, the C-3 plants. C-4 plants are characteristic plants of hot and arid environments and, since many are also important crop plants, e.g. maize and sugar cane, understanding their mode of photosynthesis is of considerable importance. In C-4 photosynthesis, of which there are several variants, the process of carbon assimilation is based partly on modified anatomy and partly on modified biochemistry. In contrast to the C-3 plant, C-4 plants have a distinctive layer of green bundle sheath cells surrounding each vascular bundle, and lying between the vascular bundle and the other green cells of the mesophyll layer (figure 3.20). CO_2 enters the leaf through the stomata, and in the mesophyll cells is converted to a C-4 compound, oxaloacetate, which is then converted to other C-4 compounds, malate and aspartate. The initial carboxylation is mediated by the enzyme phosphoenolpyruvate carboxylase (PEP carboxylase). In the least complex examples of C-4 photosynthesis, malate is transported from the mesophyll cells to the bundle sheath cells, where it is decarboxylated to pyruvate, and the released CO_2 photoreduced in the Benson-Calvin pathway in the

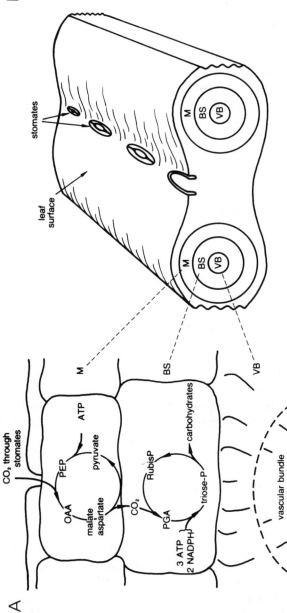

Figure 3.20 A. The pathway of carbon photoreduction in a plant possessing C-4 photosynthesis (modified from Fork (1977) in *The Science of Photobiology*, ed. K. C. Smith. Plenum).

B. A diagram of a cross-section of a leaf showing the arrangement of the tissues in a C-4 grass (from D. A. Walker (1979) *Energy, Plants and Man.* Packard Press, U.K.).

M = mesophyll cell; BS = bundle sheath cell; VB = vascular bundle.

bundle sheath chloroplast to yield triosephosphate (C-3). The pyruvate is regenerated to phosphoenolpyruvate by the enzyme-phosphopyruvate dikinase in the mesophyll chloroplast at the expense of ATP; the net effect is to raise the CO_2 concentration in the vascular-bundle sheath cells. In addition, since the PEP carboxylase has a very high affinity for CO_2, it can provide a very efficient scavenging system for CO_2 in conditions when stomata are only partially open. The individual reactions catalyzed by PEP carboxylase and phosphopyruvate dikinase are illustrated below and the integrated pathway illustrated in figure 3.20.

PHOSPHOENOLPYRUVATE CARBOXYLASE

oxaloacetate + ATP $\rightleftharpoons$ PEP + CO_2 + ADP

PHOSPHOPYRUVATE DIKINASE

ATP + pyruvate + Pi $\rightleftharpoons$ AMP + PEP + PPi

The fixation of CO_2 by PEP carboxylase and the high CO_2 concentration in vascular-bundle chloroplasts means that C-4 plants do not lose newly fixed carbon to the atmosphere, as do C-3 plants, by a process known as *photorespiration*. The mechanism of photorespiration and the source of the CO_2 released is still a matter of some conjecture, but information about the behaviour of ribulosebisphosphate carboxylase explains part of the phenomenon. Under conditions of low CO_2 and high O_2, this enzyme will behave as an oxygenase and produces phosphoglycolate in addition to phosphoglycerate.

The phosphoglycolate gives rise to other C-2 compounds, such as glycolate, glyoxylate and glycine, which themselves may be broken down releasing CO_2. In addition to the chloroplasts, the mitochondria and the peroxisomes of the leaf cell collaborate in the completion of the series of reactions associated with photorespiration, and one suggestion of the type of metabolic integration which may be involved is illustrated in figure 3.21. C-4 plants are generally regarded as having evolved from C-3 plants, as they developed an ability to utilize PEP carboxylase to reduce the loss of carbon in photorespiration. Since the Benson-Calvin pathway is the only autocatalytic cycle known to operate in photosynthesis, and the only one which results in net carbon assimilation, the PEP carboxylase system of C-4 plants can only serve as an adjunct to this cycle and not a substitute for it. In this context, the recent findings concerning the control of the synthesis of RubisP carboxylase (Fraction I protein) in the C-4 plant, maize, are of considerable interest. It appears that the code for the synthesis of the large subunit of this enzyme is present in the DNA of both the bundle sheath and mesophyll cell chloroplasts of C-4 plants, but that only in the

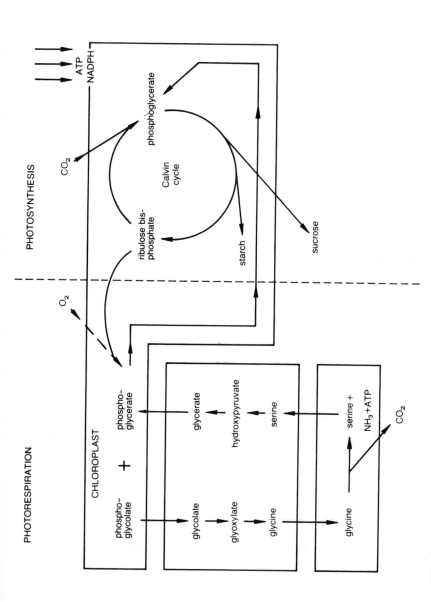

Figure 3.21 A scheme showing the interactions between photosynthesis and photorespiration in C-3 plants (from Keys (1979) *The School Science Review*, **60**, 670–78).

bundle sheath cell is the mRNA present and the large subunit synthesized (Link *et al.*, 1978).

3.5 The chloroplast in its cellular environment

The chloroplast is surrounded by an envelope composed of two continuous but quite distinct membranes, the inner of which exhibits several specific translocase properties. No function has yet been assigned to the outer envelope membrane. Purified suspensions containing fragments of both chloroplast envelope membranes have a high content of phosphatidyl choline, some unique polypeptides distinct from those in the thylakoids, and show high activity for UDP-transgalactosylase (Douce, 1974).

The molecular interface surrounding the chloroplast is of prime importance as the semi-permeable barrier allowing both regulation of chloroplast function inside the envelope membrane and regulation from outside by changes in its cellular environment. The role of the transenvelope pH gradient is one example of such regulation. On illumination, a pH gradient is established across the inner envelope membrane, protons move out of the chloroplast and K^+ moves in. The resultant alkalinization of the stroma has a direct effect on the operation of the Calvin cycle, since certain enzymes in this pathway are highly sensitive to pH. Both the activity and the activation of fructose bisphosphatase, for example, are enhanced at high pH and inhibited at acid stroma pHs; the activity of sedoheptulose bisphosphatase has an optimum pH of *c.* 8·0. An artificial lowering of stroma pH by the addition of nitrite or acetate ions to isolated intact chloroplasts severely reduces, and finally totally inhibits, CO_2 fixation. There seems no reason to doubt that such controls operate *in vivo*. The initiation of CO_2 fixation by the activation of ribulosebisphosphate carboxylase requires Mg^{++} ions, but there are conflicting reports about whether light actually stimulates Mg^{++} ion transport into the chloroplast.

The fluxes of nutrients across the envelope membrane and the movement of the products of carbon metabolism outwards are additional examples of the type of regulatory controls which exist. The main fluxes of nutrient across the chloroplast envelope involve CO_2, inorganic phosphate, triosephosphate, Mg^{++} ions and H^+ ions. The inner chloroplast envelope membrane is permeable to CO_2 which moves by unrestricted physical diffusion into the chloroplast, although some control is exerted by the stromal concentration of CO_2. Impermeability of the inner membrane to the products of carbon fixation, however, prevents depletion of the intermediates and allows photosynthesis to continue. The inner envelope membrane is almost completely impermeable to sucrose, and completely impermeable to pyridine nucleotides and hexose phosphates. Specific translocators in the inner envelope membrane, which operate on an anion counter exchange basis, catalyze the shuttle transfer of substrates and products of carbon metabolism. One of these which has a high transfer capacity for inorganic phosphate, DHAP, 3-PGA and glyceraldehyde-3-phosphate (G3P), has already been referred to. This translocator facilitates the rapid transfer of DHAP out of the chloroplast during active photosynthesis in counterexchange for inorganic phosphate, but can also mediate the entry of 3-PGA into the chloroplast from the cytosol. This supply of 3-PGA may be of critical importance in maintaining an adequate concentration of carbon cycle intermediates (figure 3.18) and as a source of phosphorylation power and reducing equivalents in the dark. A second translocator has also been identified which facilitates the movement of dicarboxylates, such as malate and oxaloacetate, and has a more limited capacity for facilitating the movement of amino acids, such as aspartate and glutamate. The dicarboxylate translocator can also play a role in energy exchange between the chloroplast and the cytosol. By facilitating the counterexchange of malate and oxaloacetate, the transfer of reducing equivalents across the chloroplast inner membrane is also facilitated, since malate dehydrogenases are present both inside the chloroplast and also in the cytosol (figure 3.22) (see Heber, 1974 for a review).

In plants with the C-4 pathway of carbon metabolism, the rapid transfer of metabolites within and between the green cells of the leaf is essential. These plants have recently been shown to have specific carriers located in the chloroplast envelope which mediate the movements of pyruvate and also phosphoenolpyruvate. Active research at present in progress aims at discovering transport capabilities of these chloroplasts and also the mode of transport of materials between the cells in the leaves of C-4 plants.

3.6 The cooperative function of chloroplasts in biosynthetic processes

The availability of suspensions of intact chloroplasts from C-3 plants capable of photoreducing CO_2 at rates similar to those found *in vivo* (Jensen & Bassham, 1966) has stimulated the examination of the additional metabolic attributes of the chloroplast. How many of its own constituents can it synthesize, and what contribution of partially elaborated intermediates does it make to macromolecular syntheses in other cellular organelles? How far can the chloroplast itself elaborate photoreduced CO_2

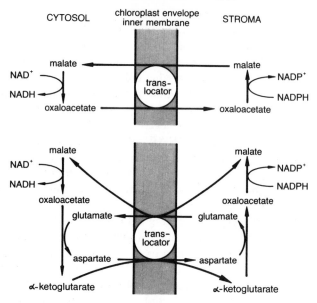

Figure 3.22   A scheme for metabolite transfer facilitated by the dicarboxylate translocation (as figure 3.18 from Heldt, 1976) in the chloroplast inner envelope membrane.

into macromolecules such as lipids, proteins and nucleic acids? As the enzyme spectrum of chloroplasts becomes established, the pattern is emerging which suggests that, while some biosynthetic pathways are initiated and completed in the chloroplast (notably the biosynthesis of pigments, nucleic acids and proteins), the source of precursors and some of the intermediates for these pathways is often outside the chloroplasts. The mature chloroplast and the other cellular components of the green cell appear to interact and collaborate in the wide variety of synthetic processes (see Givan & Harwood, 1976).

The consensus of opinion at any time on the specific intracellular enzyme locations depends on the interpretation of the results of fractionation studies, some of which are more critically carried out than others. The problems involved have been discussed elsewhere (Leech, 1978); cross-contamination of subcellular fractions and enzyme inactivation are the major ones. In the case of lipid biosynthesis and amino-acid biosynthesis in green cells, however, confirmatory reports from several laboratories now support similar interpretations of enzyme location. A summary of the pathways is given in figure 3.23 and 3.24.

The incorporation of inorganic carbon as CO_2 into organic molecules is

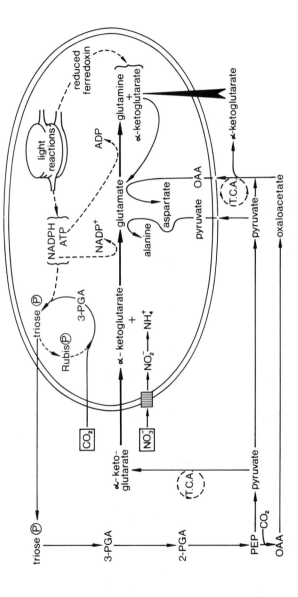

Figure 3.23 Diagram illustrating possible routes associated with the assimilation of nitrate and nitrite and the synthesis of glutamine in the chloroplast.

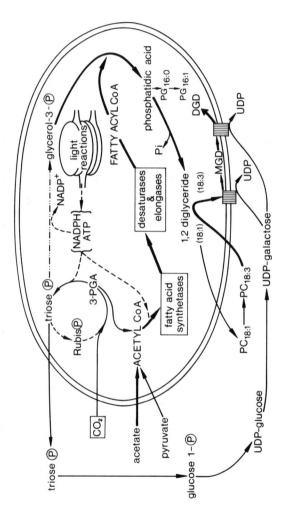

Figure 3.24 Diagram illustrating possible routes of galactolipid biosynthesis associated with the chloroplast. → demonstrated to occur in the chloroplast, – – → demonstrated in leaves, ⋯⋯→ demonstrated only in animals.

a major function of the green plant located in the chloroplast; the incorporation of inorganic nitrogen into organic nitrogen-containing molecules is a second exceedingly important one. The cyclic series of events taking place in the chloroplast is illustrated in figure 3.23. Nitrogen is taken into plants mainly in the form of nitrate ions, which are reduced by the enzymes nitrate reductase and nitrite reductase to ammonium. Nitrate reduction in leaves has long been known to be light-dependent, and the amino groups produced to be incorporated into amino acids. During photosynthetic $^{14}CO_2$ reduction, amino acids such as aspartate and alanine and phosphoserine are amongst the first products to become immediately labelled with ^{14}C; glutamate and glutamine accumulate ^{14}C-label subsequently. Both the light-dependence of nitrate reduction and the utilization of photosynthetically produced carbon chains into amino acid implicates the chloroplasts in the process of amino-acid synthesis. Considerable efforts have been devoted to assign correctly the intracellular location of nitrate reductase, and conflicting results have been obtained. Almost certainly the enzyme is rather loosely associated with the chloroplast envelope membrane (and has been demonstrated to be present in some envelope fractions), and occasionally is lost or inactivated during isolation. Rathnam (1978) has recently shown, using isolated protoplasts, that provided a continuous supply of NADPH is accessible to the chloroplast, nitrate reduction occurs there. There is no doubt that the further reduction of nitrite to ammonium (catalyzed by nitrite reductase) occurs in the chloroplast, using reduced ferredoxin as a hydrogen source. This is one reason why nitrate reduction is light-dependent. The spinach nitrite reductase has a molecular weight of 61 kD, its prosthetic group is sirohaem and there are 3 Fe atoms and 2 S atoms per molecule.

The most rapid mechanism of ammonium incorporation so far identified in intact chloroplasts is into glutamine via glutamine synthetase. This conversion occurs actively in chloroplasts in the light but not in the dark and K_m of the enzyme for ammonium is $10^{-5}M$. Two different enzyme systems capable of aminating α-oxoglutarate to glutamate, glutamic dehydrogenase (NADPH-dependent) and glutamate synthase are present in chloroplasts. Since the conversion of α-oxoglutarate to glutamate in chloroplasts is not affected by ammonium ion but is stimulated manyfold by glutamine in intact chloroplasts, it has been suggested that this is the more likely pathway to operate, although energetically it is more demanding (Lea & Miflin, 1974). α-oxoglutarate can be endogeneously generated within the chloroplast from 3-PGA since all the enzymes required, including pyruvic dehydrogenase, have recently been shown to be present in the chloroplast. Further synthesis of additional amino acids requires aminotransferases, but only two operating with glutamate (in the synthesis of aspartate and alanine) have been clearly identified in the chloroplast. There are other aminotransferases operating between aspartate and alanine, and numerous keto acids. It is possible that some keto acids are supplied from outside the chloroplast and some amino acids also, but there are inadequate data at the present time to be categorical about this. Chloroplasts have the ability to reduce sulphate to sulphide. As already mentioned, the entire machinery for the incorporation of amino acids into polypeptides is present in the chloroplast, and the first product of *in vitro* chloroplast protein synthesis to be identified was the large subunit of Fraction I protein (Blair and Ellis, 1973). Tryptic peptide analysis showed that the ^{35}S-methionine-containing peptides of the large subunit labelled *in vitro* were identical with those labelled *in vivo*. The large subunit is also synthesized when mRNA from chloroplasts is used in the wheat germ translation system.

Another example of the cooperative function of the chloroplast in metabolic synthesis is its involvement in leaf lipid biosynthesis. Illuminated isolated chloroplasts fed with acetate synthesized fatty acids at high rates, and it has been suggested that most of the fatty-acid synthesis of the leaf cell may occur in the chloroplast. The process requires oxygen and involves an acyl carrier protein which has been isolated and purified. The cellular source of acetate is unknown and, indeed, it may not be the endogeneous substrate for chloroplast fatty-acid biosynthesis. Recently it has been shown that bicarbonate is an equally good substrate, and the incorporation of ^{14}C from bicarbonate into monogalactolipid has been demonstrated. All the enzymes for the biosynthesis of acetyl CoA from 3-PGA, including the pyruvic dehydrogenase complex, are present within the chloroplast. As with acetate, however, chloroplasts fed with bicarbonate only synthesize a limited spectrum of fatty acids, predominantly stearate and oleate, and despite numerous attempts only traces of the major trienoic fatty acid of chloroplasts (α-linolenic acid) have ever been shown to be synthesized by chloroplasts *in vitro*. Alternative suggestions for its synthesis have therefore been made which implicate phosphatidyl choline in an acyl carrier role. This suggestion is based on the results of experiments in which leaves are fed with $^{14}CO_2$ followed by a cold chase. In these experiments ^{14}C becomes incorporated at the 2-position of phosphatidyl choline (PC) as oleate, subsequently linoleate appears and finally linolenate, only later still are galactolipids labelled in linolenate recovered. It is suggested that diglycerides, containing predominantly oleate, leave the chloroplast and become incorporated into cytoplasmic PC. Desaturation occurs in the PC molecule, and unsaturated acyl groups are transferred onto galactose residues during galactolipid synthesis in the chloroplast envelope. Several indirect lines of evidence support this suggestion which is shown pictorially in figure 3.24. Transacylation between PC and diglycerides has never been shown to occur directly in the chloroplast.

The modern chloroplast, whatever its origin and potential in its evolutionary past, is dependent on its cell environment for survival. Both in its development and metabolic syntheses, and in the maintenance of its ability to photosynthesize, it is involved in sophisticated interactions with other cellular components. The mature chloroplast may survive for limited periods in an artificial environment, but it seems highly unlikely that the whole plastid life cycle could be completed in such conditions.

CHAPTER FOUR

MITOCHONDRIA

4.1 Structure and distribution of mitochondria

One of the functions of science is to bring order out of chaos. At the end of the nineteenth century cyto-chaos reigned, at least in the mind of the observer. For fifty years almost every cytologist of note had reported granular thread-like particles in the cytoplasm of aerobic cells, and assigned a name and function to them—fila, chondriokonts, fädenkörner, blepharoblasts, vermicules and others. As the twentieth century progressed and a basic uniformity of function became apparent, the name *mitochondrion* (Benda, 1898) became generally accepted. It is now clear that, although mitochondria in different cell types may vary in biochemical details and in size, structure and frequency, they all contain the enzymes of the tricarboxylic acid cycle (TCA cycle) and carry out oxidative phosphorylation—ATP synthesis coupled to substrate oxidation. They also all conform to a basic structural pattern, namely an outer membrane or envelope enclosing an inner membrane which has tube-like invaginations (cristae) into an inner compartment (the matrix). The cristae membrane of liver mitochondria (figure 4.1) is 3–4 times greater than the outer membrane area. Some mitochondria, particularly those from kidney, heart and skeletal muscle (figures 4.2 and 4.3), have more extensive cristae arrangements than liver mitochondria, while others (e.g. from fibroblasts, nerve axons and most plant tissues) have relatively few cristae. Mitochondria in epithelial cells of carotid bodies have only four or five cristae, and mitochondria from non-myelinated axons of rabbit brain have only a single crista. The cristae membranes are the sites of certain TCA cycle enzymes, respiratory chain components, and the enzyme, ATP synthetase; therefore, not surprisingly there is a rough correlation between the respiratory rates of mitochondria and the extent of their cristae. Typically tissues with high respiration rates have mitochondria with many cristae.

Estimation of the number and size of mitochondria in cells requires serial sectioning techniques and is sometimes complicated by the propensity of

82

mitochondria to fuse and separate. In certain Euglenoid cells, the mitochondria fuse into a reticulate structure during the day and dissociate during darkness. Similar changes have been reported in yeast species, apparently in response to culture conditions. Little is known about fusion and separation in higher cells but, if it exists, it is less dramatic than in these lower eukaryotic cells. Neither the biological significance nor the molecular mechanisms of fusion has been defined. The number of mitochondria per cell varies from only one in certain haemoflagellate species to several hundred thousand in sea urchin ova and giant amoebae. The question of the selection pressures that have influenced the number, size and structure of mitochondria is intriguing. What is the significance of hepatic cells having some 1500 mitochondria sparsely endowed with cristae rather than 300 with many cristae? The answer may lie partly in the type of work that mitochondria power. Typically mitochondria with many cristae are

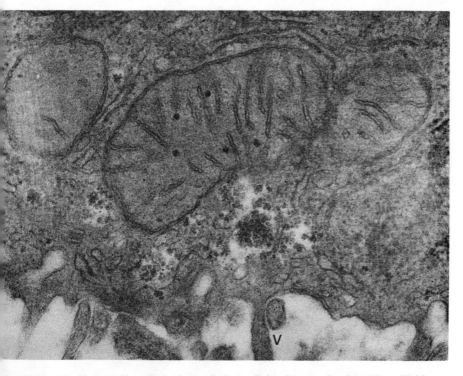

Figure 4.1   Electron micrograph of part of a liver cell showing mitochondria. Microvilli (v) extend into the bile canaliculi (P. G. Humpherson) (× 54 000).

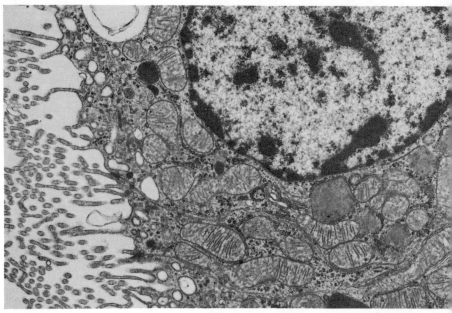

Figure 4.2 Electron micrograph showing numerous mitochondria with extensive cristae found in proximal tubule cells of rat kidney. (Dr. J. Baker) (× 13 300).

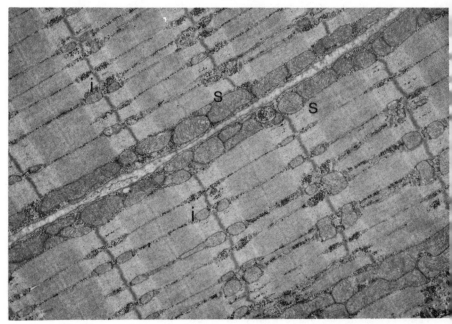

Figure 4.3 Electron micrograph of skeletal muscle from rat hind limb showing the regular arrangement of heavily cristaed mitochondria lining the sarcolemma (subsarcolemmal mitochondria—s) and between the muscle fibrils (interfibrillar mitochondria—i).
(Dr. Gillian Bullock) (× 10 000).

associated with mechanical and osmotic work situations where there are sustained demands for ATP and where space is at a premium, e.g. between muscle fibres and in the basal infoldings of kidney tubule cells. Since the work of hepatic cells is mainly biosynthetic and degradative, and the work locations are spread throughout the cell, in these cells it may be more efficient to have a large number of "low key" sources of ATP production distributed throughout the cytosol. There are many instances of mitochondria seemingly located strategically at work sites, for example, in the oocyte of *Thyone briaeus*, rows of mitochondria are closely associated with endoplasmic reticulum membranes, where ATP is required for protein synthesis. Mitochondria are often particularly numerous in regions where ATP-driven osmotic work occurs—the brush border of kidney proximal tubules, the infoldings of the plasma membranes of dogfish salt glands, the Malpighian tubules of insects, the contractile vacuoles of some protozoans. The mechanisms that partition mitochondria into particular intracellular regions have not been seriously investigated. Do outer membrane components "lock on" to specific cytoskeletal structures or membranes by ionic interaction? Is mitochondrial distribution effected by different cytosol viscosities? These and other questions are open to experimental attack.

It would be wrong to assume that demand for ATP is the only, or even the major, determinant of mitochondrial structure and distribution. In brown adipose-tissue mitochondria, the oxidative capacity is greater than the potential for ATP synthesis, which correlates with other evidence that these mitochondria are specialized for producing heat to maintain body temperature. Non-myelinated axons contain many mitochondria that are poor ATP factories, since each has only a single crista. Perhaps the dominant selection factor in this case is a requirement for monoamine oxidase, an enzyme on the outer mitochondrial membrane that oxidatively deaminates monoamines, including neurotransmitters, maximum outer membrane area being achieved by force of numbers. Mitochondria may also be important as ion buffering systems that regulate aspects of metabolism. The total intramitochondrial volume in some cell types can approach 25% of the cytosol volume and, given the ability of mitochondria to take up and release Ca^{2+} and other regulatory ions, they could influence significantly the rates of ion-dependent processes. There is, in fact, evidence for a mitochondrial role in the Ca^{2+}-dependent rhythmic contraction and relaxation of heart and uterine muscle.

Typically mitochondria are spherical or rod-shaped. However, in some cases, the shape is markedly adapted for specific cell situations, e.g. in the

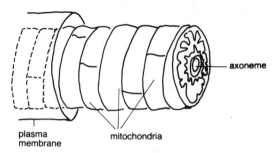

Figure 4.4   Diagrammatic representation of the proximal midpiece of mammalian sperm showing the packing of mitochondria round the axoneme.

midpieces of mammalian spermatozoa, the mitochondria are in the form of crescentic rods that pack helically round the axoneme (figure 4.4). Presumably this gives a high ATP-producing capacity for weight ratio, a useful feature in motile cells. Variations on this theme are found in earthworm sperm, where mitochondria pack longitudinally along the midpiece, and in sea urchin sperm where annular mitochondria encircle the axoneme. Branched mitochondria are frequently found wrapped round the myofibrils in muscle tissue, and seem to strengthen the tissue mechanically, as well as provide ATP for contraction. An unusual form occurs in bovine Graafian follicles, where some mitochondria have an end pulled out to form a hood that encloses a segment of endoplasmic reticulum, thereby anchoring the mitochondria in the protein-synthesizing region. Although we can see some rhyme and reason in the diverse morphology and distribution patterns of mitochondria, very little is known about the factors that influence their morphogenesis.

4.2 Cellular origin of mitochondria

The main problem in determining the cellular origin of mitochondria is that their small size precludes accurate analysis by time-lapse phase-contrast microscopy. However, E.M. and tracer experiments indicate that the mitochondrial complement in most cells is maintained by growth and division of existing mitochondria. Binary fission has been clearly shown in the single mitochondrion of the flagellate *Chromulina,* and septa or budding have been reported in mitochondria of several cell types. A key experiment by Luck (1963) involved growing a choline-requiring mutant of *Neurspora* in medium containing ^{3}H choline, and following the distribution of the ^{3}H label in mitochondria during subsequent growth in medium containing cold choline. Choline is incorporated into the phospholipid of mitochondrial membranes, and the rationale was that, if mitochondria arose *de novo,* the ones synthesized in the presence of ^{3}H choline would be labelled and those synthesized after replacement of the label would be unlabelled. In fact, the label was distributed throughout the mitochondrial complement. Controls suggested that this was not due to turnover or exchange of ^{3}H choline and cold choline, and the most reasonable interpretation is that mitochondria arise by incorporation of material into existing mitochondria and the subsequent division of these mitochondria. However, the evidence does not exclude the possibility that the labelling patterns were due to fusion between labelled mitochondria and newly synthesized cold mitochondria, followed by random fragmentation.

De novo assembly of mitochondria from bits and pieces has frequently been proposed, but much of the earlier evidence for this has proved to be artefactual. When the facultative anaerobe *Saccharomyces cerevisiae* is grown anaerobically, functional mitochondria and certain aerobic dehydrogenases and cytochromes disappear after some hours, the normal situation being restored only after admission of air to the culture. However, membranous structures recognizable as deficient mitochondria are present, even during the repressed phase, and provide a scaffolding into which the missing components are assembled after the repression is lifted. Although the possibility of *de novo* synthesis, or even the genesis of mitochondria from other internal membranes, cannot be discounted, the evidence is weak and the biological advantages are not immediately apparent. In the next section we will see that mitochondria have a sophisticated genetic system, the continuity of which would be better safeguarded if mitochondrial biogenesis was achieved by growth and division of pre-existing mitochondria.

4.3 DNA replication and protein synthesis

In 1963 Nass and Nass demonstrated fibres of the size order of DNA in the matrix of mitochondria of embryonic cells. On treating the sections with DNAse, the fibres were degraded, suggesting that they were DNA. This was followed by various reports that firmly established the reality of mitochondrial DNA, including evidence that mitochondria could incorporate ^{3}H thymidine into DNA, and that the DNA isolated from mitochondrial preparations had different base ratios and buoyant densities from nuclear DNA. Mt DNA varies in length from about 5 μm for most animal species to 30 μm or so in higher plants. There is a range of sizes in fungi, *Schizosaccharomyces pombe* having mt DNA of 6 μm, while species of *Saccharomyces* and *Neurospora* have mt DNA of 25 μm. At 18 μm the DNA of the protozoans *Paramecium* and *Tetrahymena* is about the middle of the range. Mt DNA is double-stranded and covalently circular in most cases, often appearing as twisted supercoils. The mechanism of replication is poorly understood, but a model that fits many of the data for animal cells is shown in figure 4.5. The strands separate at a point, and replication of one strand occurs unidirectionally, almost to completion before replication of the other strand starts. The model is based on E.M. studies, and the isolation and analysis of intermediates. The mitochondrial DNA polymerase involved in the synthesis is a different species from nuclear polymerases.

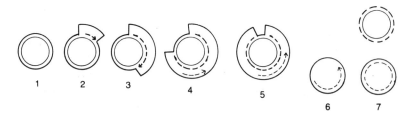

Figure 4.5   Model of replication of mt DNA.
Parental strands = solid lines; progeny strands = dashed lines; heavy DNA strands = thick lines; light DNA strands = thin lines. Proposed by Robberson *et al.* (1972) *Proc. Natl. Acad. Sci. U.S.* **69**, 738.

Mitochondria have been reported to contain from 1 copy (yeast) to 10 copies (vertebrate cells) of mt DNA. However, if mitochondrial biogenesis in yeast is repressed by high glucose in the growth medium, there is a substantial increase in the mt DNA per mitochondrion. In fact, mt DNA appears to be maintained in a constant ratio to n DNA, suggesting that it is under nuclear control. Experiments with synchronized cells indicate that in some cells, namely mammalian cells in culture, mt DNA is synthesized at specific times in the cell cycle (S and G2), whereas in *Tetrahymena* and *Physarum* it occurs continuously. Therefore different controls may operate in different cell types. Possibly programmed transcription of a nuclear gene for an initiator of mt DNA synthesis determines the kinetics of replication, but the dearth of fact makes model building at this stage an academic exercise.

Mitochondria possess the complete molecular machinery for the transcription of DNA and the synthesis of protein—DNA-dependent RNA polymerase, ribosomes (mitoribosomes), all the necessary species of tRNA, amino acid—tRNA synthetases and factors for initiation, elongation and termination of protein synthesis. All these proteins and mt RNA species are structurally different from their counterparts in the cytosol. Mitoribosomes conform to the usual two-subunit structure, but differ from cytoribosomes and bacterial ribosomes in size, composition and molecular weight. With S values between 72 and 76, depending on species, they are intermediate between the 80S cytoribosomes and the 70S bacterial ribosomes. The DNA replicating and transcribing system in mitochondria and the machinery for protein synthesis are therefore basically similar to those in the nucleus and cytosol, but distinct and different in detail.

In chapter 3 it was noted that the chloroplast possessed biosynthetic capacity, but only for a limited number of its constituents. What then is the role of the mt biosynthetic systems, and to what extent are mitochondria autonomous organelles? It has been more difficult to identify protein and RNA molecules synthesized by isolated mitochondria than by isolated chloroplasts, because of lower activity in the former. However, the differential effect of inhibitors of protein synthesis on cytoribosomes and organelle ribosomes is very sharp in the case of mitoribosomes, and this is the approach that has been most widely applied. Cells are incubated with radioactive amino acids in the presence of chloramphenicol or spiramycin, which binds to the large 50S subunit of mitoribosomes, or with cycloheximide which acts specifically on the large 60S subunits of cytoribosomes. Protein synthesized when cycloheximide is present is regarded as the product of mitoribosome activity. The least equivocal results have been obtained from studies on yeast and *Neurospora,* but the method has been applied to tissue macerates and mitochondria *in vitro* in short-term experiments using the incorporation of amino acids of high specific radioactivity as a measure of protein synthesis. Other methods used include investigation of the patterns of protein synthesis in yeast mutants that are deficient in mitochondrial protein synthesis. Work in several laboratories has shown that the main products of mitochondrial protein synthesis in the yeast *S. cerevisiae* are:

1. *The three largest subunits* (MW 40 000, 27 000, 25 000) *of cytochrome oxidase (a-a₃)*, the terminal oxidase of the respiratory chain. The four smaller subunits (13 800, 13 000, 10 200, 9500) of this oxidase are synthesized on cytoribosomes.
2. *The apoprotein of cytochrome b* (MW 30 000).
3. *Four subunits of the membrane factor of ATP synthetase.* ATP synthetase is composed of

nine or more different subunits organized in two dissociable complexes—F_1 and membrane factor (figure 4.20). F_1 is hydrophilic, has five subunits and contains the binding sites for ADP, Pi and ATP. It is associated with a complex of at least four different subunits, collectively called *membrane factor* (F_0), as it is sited deep in the inner membrane. In most tissues F_1 is attached to membrane factor by a protein (OSCP). F_1 and OSCP are synthesized on cytoribosomes but the membrane factor proteins are synthesized in mitochondria.

4. *One protein component of the many that compose the small mitoribosome unit.*

This is possibly not a complete list, for small numbers of low-MW proteins are not easily detected. It should not be regarded as rigidly applying to all species. In *Neurospora* at least one of the membrane factor proteins is synthesized on cytoribosomes. The available data for animal mitochondria broadly agrees with the results for yeast. The very limited number of proteins synthesized in mitochondria shows that these organelles have no significant autonomy.

4.4 Mapping the mitochondrial genome

The positions of over 20 genes have now been mapped on the mt DNA of *S. cerevisiae*. This has been made possible by the isolation and genetic analysis of mutants showing respiratory deficiencies due to damage to their mitochondrial genome. In most cell types, a mutation causing loss of respiratory function would be lethal, but yeast cells are facultative anaerobes able to obtain enough energy for growth from the fermentation of glucose to ethanol. Thus they can survive and grow independently of oxidative phosphorylation if a fermentable sugar is present in the medium. The first respiratory-deficient mutants were discovered by Ephrussi in 1949 and termed *petite* mutants because they formed small slow-growing colonies in contrast to wild-type cells. Petite mutants occur in nature with a relatively high frequency ($0\cdot1-1\cdot0\%$) and it is possible to convert almost 100% of a wild-type population into petites by adding ethidium bromide, a DNA-intercalating agent. They have ill-defined membranous structures instead of mitochondria, and spectroscopic analysis shows that the respiratory chain cytochromes *b* and *a-a$_3$* are missing.

The life cycle of *S. cerevisiae* makes it a good organism for genetic analysis. Mating occurs between haploid mating types to produce zygotes that give rise to diploid cells by budding. On sporulation, the diploid cells undergo meiosis to give haploid ascospores. Therefore the inheritance of morphological or metabolic features can readily be followed and quantified. Using methods that are well described by Mahler (1973), Ephrussi found that, unlike mutations of nuclear genes, most petite mutations showed non-Mendelian inheritance. This indicated that the genetic determinant for petite conditions was cytoplasmic. This was later confirmed by the finding that petite mutants of this class had either no mt DNA (ρ^0 petites) or grossly altered DNA (ρ^- petites). In the latter case the mt DNA had suffered substantial deletion, and the remaining mt DNA had amplified or repeated itself to approximately the same total mt DNA as in the wild type. Petite mutants cannot carry out protein synthesis and never revert to the wild-type phenotype—not surprising considering the massive damage to the mt DNA.

Petites are useful but blunt tools for genetic experiments, and have been used in conjunction with other mutants with mitochondrial lesions. These show cytoplasmic inheritance, but have no gross damage to the mt DNA and are considered to be the result of point mutations, as reversion to the wild type is not uncommon. They include mutants that are resistant to drugs that affect mitochondrial functions, namely oligomycin (OLI), chloramphenicol (C), erythromycin (E) and paramomycin (P), and a series of mutants that lack cytochrome b or cytochrome oxidase or both. There is therefore a wide spectrum of mitochondrial mutants ranging from ρ^0 petites without mt DNA, through the ρ^- mutants with various degrees of deletion to the mit$^-$ mutants with point mutations, in all a useful battery for genetic analysis.

The most widely used technique for mapping mt genes is recombination frequency. In brief, if two compatible mating strains carrying, for example, resistance to the drugs erythromycin and chloramphenical respectively are crossed, a proportion of the progeny will be stable strains carrying resistance to both drugs. How mt DNA molecules in different mitochondria interact with one another to produce recombinants is poorly understood. At the time of nuclear fusion during zygote formation, the mitochondrial membranes become disorganized. This is thought to facilitate fusion of mitochondria and mixing of their DNA, leading to crossover and segment exchange, giving recombinant mt DNA which is randomly assorted into mitochondria, and subsequently into the daughter cells produced by budding. The principles of linkage mapping are the same as for the bacterial genome—the greater the recombination frequency of genetic markers, the further apart they are on the chromosome. Deletion mapping is also frequently employed. Without going into detail, this estimates the degree of co-retention of markers after deletion of mt DNA, the premise being that the closer they are together on the DNA the more likely genes are to be lost or retained together.

The linkage map of mt genes of *S. cerevisiae* shown in figure 4.6 is compiled from experiments done in several laboratories. The sites of the

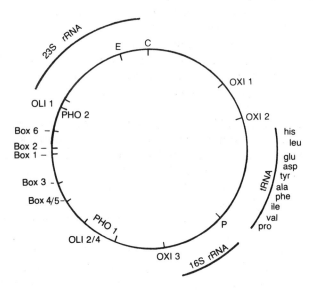

Figure 4.6   Genetic map of the mt DNA of *Saccharomyces cerevisiae*.
C = chloramphenicol resistance; E = erythromycin resistance; P = paramomycin resistance;
OLI 1 and 2/4 = oligomycin resistance; BOX 1–6 = cytochrome *b* deficiency; OXI 1, 2 and
3 = cytochrome oxidase markers.

mutations that conferred drug resistance were among the first to be
determined and served as important reference points for mapping other
sites. Mutants that show resistance to inhibition of the ATP synthetase by
oligomycin map in two regions, OLI 1 - PHO 2 and OLI 2 - PHO 1 -OLI 4.
Oligomycin acts at the level of the membrane factor of the enzyme, and
it is reasonable to think that the two mutation regions mark the genes for
at least two of the four known subunits. In fact OLI 1 and PHO 2 are
markers for a proteolipid subunit of membrane factor that binds
dicyclohexylcarbodiimide (DCCD), an inhibitor of ATP synthetase that is
functionally similar to oligomycin. The gene for proteolipid has recently
been sequenced by Macino and Tzagoloff (1979) using petite mutants from
which other gene markers had been deleted. The nucleotide sequence
consists of 76 codons that confirm the known primary structure of
proteolipid (76 amino acids). The gene has some interesting features,
including a high selectivity of codons for amino acids, for example,
although the amino acid leucine has six possible codons, 11 out of the 12

leucines in the proteolipid are coded by just one codon (UUA). Similarly only one of the six possible codons for serine is used. The significance of this is not clear, but it suggests that there is little degeneracy in the mt code. Other regions of the mt genome are currently being sequenced and the results will clarify this point. The mutations mapping at OLI 2, PHO 1 and OLI 4 have not yet been identified with specific gene products. They may well mark different regions of a single gene coding for a single membrane factor subunit.

The 16S and 23S rRNA species that are present in mitoribosomes are coded by mt DNA, and their loci were mapped by looking at their hybridization affinities with DNA from petites known to have retained particular segments of mt DNA; for example, it was found that 23S rRNA hybridized strongly with DNA extracted from petites that carried the E and C drug resistance loci, but not OLI 1 or OLI 2. One of the methods used to find the loci for the tRNA species was to measure their degree of hybridization with defined fragments of mt DNA obtained by restriction endonuclease cleavage. Slonimski and co-workers investigated some 90 mutants deficient in cytochrome oxidase and found that they could be assigned to only three loci, which were named OXI 1, 2 and 3. Biochemical analysis of the mutants showed that those with mutations that mapped at OXI 1 lacked subunit 2 of the oxidase, while those mutations mapping at OXI 2 and 3 had lost subunits 3 and 1 respectively. This fits beautifully with the finding mentioned in section 4.3, that subunits 1, 2 and 3 of the oxidase are synthesized in mitochondria.

Possibly the most interesting result in recent years relates to cytochrome *b*. Fine structure mapping was carried out by Slonimski and co-workers on over twenty mutants deficient in cytochrome *b*. When the mutation for each mutant was mapped, it was found that they fell into five groups separated by long stretches of DNA. Each group was considered to mark a piece of DNA that coded for part of the cytochrome *b* or some factor important for its expression. These groups, designated Box loci, are shown in sequence on the linkage map (the numbering is based on the chronology of discovery). A detailed study of the Box region using different techniques has shown that the genetic information for the polypeptide resides mainly at Box loci 6, 1 and 4/5. Cytochrome *b* is a single polypeptide of 30 000 MW and would require only about 900 base pairs to code for it. However, as shown below, the pieces of the gene are spread over a segment of DNA that could be up to 8000 or so base pairs long.

Box 6		Box 1		Box 4/5
100–300	1800–3000	100–400	3200–4500	400–500 bp

Two clusters of mutational sites Box 2 and Box 3 map between the parts of the split cytochrome *b* structural gene in the order 6-*2*-1-*3*-4/5. These loci are of the same size order as the Box 6, 1 and 4/5 loci, and their functions are not clear. Slonimski *et al.* (1978) have postulated that they may carry information that is important for the processing of the primary RNA transcript of the region, e.g. the nicking out and ligation of the coding sequences for cytochrome *b* into functional mRNA; or control the expression of cytochrome oxidase.

Recent work has shown that the number and length of inserted sequences in the cytochrome *b* gene is strain-dependent. The mt 23S RNA gene in yeast and *Neurospora* is also "in pieces", so the phenomenon of split genes discussed earlier for nuclear DNA evidently applies to at least some organelle genes.

Since mt DNA in *S. cerevisiae* is 25 μm long compared with 5 μm in animal cells, the above results should not be extrapolated to other species. Only 15 000 base pairs are available on 5 μm of DNA, and the combined molecular weights of the proteins coded by yeast mt DNA, together with the mt t and rRNA species, would require something over 13 000 base pairs. Therefore the genes in animal mitochondria are likely to be closer together and the cytochrome *b* gene can scarcely be spread over 8000 or so base pairs as in the aforementioned yeast strain. In fact, some genes in yeast mt DNA are absent from mt DNA in higher cells. *Tetrahymena* mitochondria synthesize tRNA for only four amino acids, compared with the thirty tRNA species in yeast mt DNA, the remainder being synthesized in the cytosol. In mammalian HeLa cells, 12 mt tRNAs hybridize with the heavy mt DNA strand and 5 with the light strand, indicating that both strands are transcribed—an interesting way to accommodate more genes in limited DNA. Whether genes on both strands is a common strategy for animal mt DNA remains to be seen. Mammalian mt DNA cannot be mapped by mating studies, but attempts are being made to identify cloned restriction fragments with specific gene products and elucidate gene sites and coding principles. Informed sources say that surprises are on the way.

4.5 The assembly of mitochondria

From the foregoing, it is evident that the assembly and maintenance of mitochondria requires the co-operation of two systems of genetic expression, the mitochondrial and the nuclear. The contributions of the two genetic systems to the functional organelle in *S. cerevisiae* are outlined in figure 4.7. The structural and catalytic proteins responsible for the basic framework of the mitochondrion are under nuclear control and synthesized in the cytosol, the genetic evidence for this being supported by the fact that mitochondrial-like structures are formed even in ρ^0 petites which are devoid of mt DNA. The hundred or so proteins (including mt ribosomal proteins) necessary for the mt transcription and translation systems must be imported from the cytosol into the matrix. A large number of enzymes involved with the TCA and β oxidation cycles must also be brought in from outside. The problem of how mitochondria recognize and import proteins is still not elucidated. One hypothesis is that cytoribosomes bearing the appropriate mRNA bind at specific sites on the outer

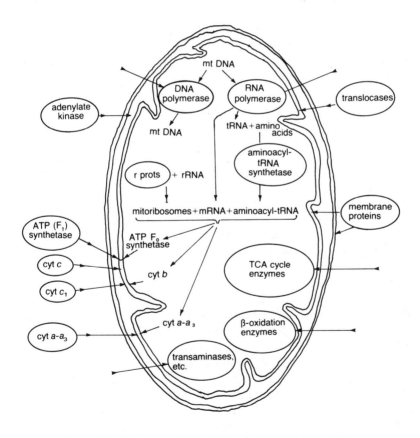

Figure 4.7   A summary of the origins of mitochondrial proteins.

membrane, where it is in contact with the inner membrane, and discharge protein into or through the membranes as the protein is synthesized. This "direct insertion" model is supported by the fact that regularly-arrayed tightly-bound cytoribosomes are frequently seen on the outer membranes of yeast mitochondria. Under suitable *in vitro* conditions, the bound ribosomes incorporate radioactive amino acids into proteins which appear to be inserted into the membrane rather than released to the medium as they are largely immune from degradation by added proteases unless the membrane structure is disrupted by deoxycholate. However, the general applicability of this model is in doubt. Membrane-bound cytoribosomes are not widely evidenced outside yeast mitochondria, and tight coupling between protein synthesis and transport across the membrane is not always observed.

Recently there has been a strong move towards the view that proteins destined for mitochondria are synthesized on free cytoribosomes as larger precursors with "transit" peptides at one or both ends which bind to a specific receptor on the target membrane. Binding would induce the opening of a polar transmembrane duct down which the protein would diffuse. The "transit" sequence would be cleaved when the protein arrived at its destination in the membrane or the matrix. In support of this, recent work by Maccecchini *et al.* (1979) has shown that the three large subunits of the mitochondrial F_1 ATP synthetase are synthesized in the cytosol as precursors of molecular weight 4% to 15% greater than the mature subunits. These precursors are imported into the mitochondria being processed to mature subunits during or after translocation. Mature F_1 subunits are unable to permeate the membrane, supporting the view that the precursors have special transport properties. Although the proteolipid of ATP synthetase is synthesized in yeast mitochondria, it is nuclear-coded and synthesized on cytoribosomes in *Neurospora crassa*. Sebald's group have recently identified a precursor form of the *Neurospora* proteolipid (MW 12 000 compared with 8000 for the mature proteolipid) and suggest that this form may have a transit peptide. In view of the propensity of hydrophobic proteolipid for membranes in general, it would seem eminently reasonable for it to be produced in the form of a precursor with recognition properties for mitochondrial membranes. Several other recent papers encourage the transit peptide view. Cote *et al.* (1979) have shown that one and possibly three of the subunits of cytochrome b-c_1 complex first appear in the cytosol in the form of larger subunits synthesized on cytoribosomes. Either before or after trimming these subunits must enter the inner membrane and link up with the cytochrome b apoprotein synthesized on mitoribosomes. Cytochrome oxidase, the other membrane component compiled from the products of both cytosol and mitochondrial protein synthesis, has a neat way of bringing its four cytosol synthesized subunits to the inner mitochondrial membrane. They are linked together on one large precursor and apparently separate only when they reach their site. This implies that their genes are contiguous and that they are co-transcribed and co-translated. In the case of proteins that are mosaics of subunits from both cytosol and mitochondria, it seems that the dominant control is exercised by the nucleus through cytosol translation products. An account of some relevent experiments can be found in Tzagoloff et al. (1979).

Several cytosol-synthesized enzymes that operate in the mitochondrial matrix are being investigated to determine whether they too have precursor forms that facilitate their transport. A unifying hypothesis is not yet possible, and molecular mechanisms must await more information on protein structure and kinetics. Conceivably different classes of proteins could be inserted into and through mitochondrial membranes by different methods including cytoribosome attachment and direct insertion. Happily there is no hint of mitoribosome-synthesized proteins contributing to the outer mitochondrial membrane. This membrane is composed wholly of proteins synthesized in the cytosol.

Studies on the turnover of mitochondrial proteins have produced conflicting data, and readers are referred to a review by Mayer (1979) for guidelines on the subject. In summary, cytochrome components of the inner membrane tend to turn over co-ordinately with rather similar half-lives. The proteins of the outer membrane turn over faster than those of the inner membrane, and again there is some co-ordinated turnover of different proteins. In contrast, mitochondrial non-membrane enzymes show much faster turnover and marked heterogeneity.

The fundamental questions that lie beyond the answers to the current questions on mitochondrial assembly include why and how some proteins are replaced faster than others, how the integrity of the various multienzyme systems is maintained, and the nature of the factors that modify mitochondrial structure and function during differentiation.

4.6 Intermediary metabolism in mitochondria

Mitochondria are the sites of the terminal catabolism of the bulk nutrients—polysaccharides, lipids and proteins. The preliminary digestion of these compounds occurs in the cytosol, leading to simple sugars, fatty acids, glycerol and amino acids. Sugars and glycerol are further broken down in the glycolytic enzyme sequence to pyruvate, which enters a final catabolic sequence in the mitochondrion, the tricarboxylic acid (TCA) cycle (figure 4.8). The TCA cycle enzymes are present in soluble form in the matrix or lightly bound to the inner surface of the inner membrane. An exception is succinic dehydrogenase which is membrane bound and difficult to remove. Substrates entering the cycle undergo oxidation and decarboxylation to CO_2 and water.

The TCA cycle is the end of the degradation routes for fatty acids and amino acids as well as sugars. Fatty acids are first activated, i.e. a thioester linkage is formed between the carboxyl group of the acid and the sulphydryl end of Coenzyme A. The energy is provided by ATP, and the reaction is catalyzed by acyl-CoA synthetase, which is associated with the mitochondrial outer membrane. The activated fatty acid is translocated across the inner membrane to the matrix, where it is degraded in the β-oxidation sequence (figure 4.9) to acetyl CoA fragments that enter the TCA cycle.

Excess amino acids are also processed, mainly in the cytosol, before entering the cycle. There is considerable tissue/organism variation in amino-acid degradation, but the main strategy in liver is the transfer of amino groups to α-ketoglutarate, with the formation of L-glutamate and keto acids in reactions catalyzed by transaminase enzymes, e.g.

$$\text{L-aspartate} + \alpha\text{-ketoglutarate} \rightleftharpoons \text{oxaloacetate} + \text{L-glutamate}$$
aspartate transaminase

$$\text{L-alanine} + \alpha\text{-ketoglutarate} \rightleftharpoons \text{pyruvate} + \text{L-glutamate}$$
alanine transaminase

The keto acids formed are often TCA intermediates that enter the mitochondria and are oxidized in the TCA cycle. Glutamate is also taken into the mitochondria, where it is deaminated to the TCA intermediate α-ketoglutarate and free ammonia:

$$\text{L-glutamate} + NAD^+ + H_2O \rightleftharpoons \alpha\text{-ketoglutarate} + NH_4^+ + NADH$$
glutamate dehydrogenase

This reaction is pulled to the right by the conversion of free ammonia to carbamoyl phosphate, a key intermediate in the urea cycle (figure 4.10). This compound reacts with ornithine to form citrulline, which passes into the cytosol where the rest of the urea cycle occurs. Thus liver mitochondria both oxidize the carbon skeleton of amino acids and channel their toxic amino groups into the urea cycle. The importance of this latter process in liver mitochondria is shown in that the enzyme carbamoyl phosphate synthetase is the most abundant soluble protein in the matrix, and there is a

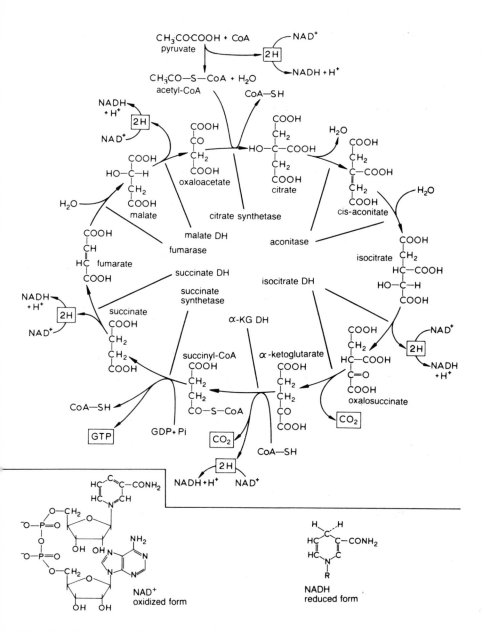

Figure 4.8 The tricarboxylic acid (TCA) cycle with the structure of the oxidized and reduced forms of nicotinamide adenine dinucleotide (inset).

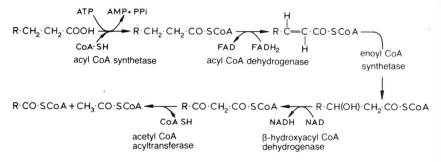

Figure 4.9 The oxidation sequence for fatty acids.

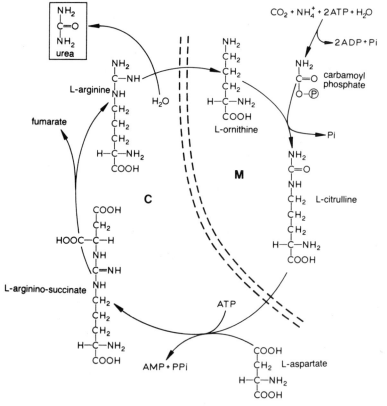

Figure 4.10 The urea cycle in liver. Note its compartmentalization between mitochondria and cytosol.

special mitochondrial membrane carrier that exchanges extramitochondrial ornithine for intramitochondrial citrulline. This is an interesting case of cooperation between two compartments—as figure 4.10 shows, the urea cycle removes free ammonia from the mitochondrion and excess amino groups from the cytosol, aspartate being the donor. Transaminase enzymes are present in the mitochondrial matrix as well as in the cytosol. Their ability to interconvert key amino acids and keto acids increases the scope and flexibility of mitochondrial metabolism, and they also play a part in the malate-aspartate shuttle (page 149) for the oxidation of cytosol NADH.

Mitochondria play a biosynthetic role, most notably the synthesis of phosphoenolpyruvate when TCA intermediates are present in excess after food, thereby assisting gluconeogenesis (sugar biosynthesis). Phosphoenolpyruvate cannot be synthesized directly from pyruvate, because pyruvate kinase is irreversible, but mitochondria accomplish this indirectly by means of a shuttle. The mitochondrial enzyme pyruvate carboxylase catalyzes the formation of oxaloacetate from pyruvate

$$\text{pyruvate} + CO_2 + ATP \xrightarrow{\text{acetyl CoA}} \text{oxaloacetate} + ADP + Pi$$

An advantage of this occurring in mitochondria is that high intramitochondrial levels of the positive modulator acetyl CoA and the cosubstrate ATP are signals that there is excess oxidizable substrate, and that the cell is in a highly charged energy state, i.e. that it can afford to convert metabolic intermediates back to sugars. In some species, the resulting oxaloacetate is converted in the mitochondria to PEP.

$$\text{oxaloacetate} + GTP \longrightarrow \text{phosphoenolpyruvate} + CO_2 + GDP$$

In others the reaction occurs in the cytosol. The cytosol contains all the enzymes for the conversion of PEP to glucose-6-phosphate, but even here the sequence is controlled by the overall energy balance of the cell. The hexosediphosphatase step is allosterically inhibited by AMP and stimulated by citrate. This is good economic sense, for high AMP levels in the cytosol mean that more ATP should be generated and that sugar breakdown rather than synthesis is required. In contrast, high citrate levels in the cytosol mean that the mitochondria have so much substrate that there has been overproduction of intermediates and that sugar synthesis should be promoted. In fact, high citrate concentrations in the cytosol also leads to fatty-acid synthesis, for citrate is cleaved by a citrate-cleaving enzyme to acetyl CoA, which is a source of carbon for fatty acids when it is found in the cytosol.

$$\text{citrate} + ATP + CoA \longrightarrow \text{acetyl CoA} + ADP + Pi + \text{oxaloacetate}$$

Since in most cases the citrate in the cytosol has been synthesized in the mitochondria and transported out because it is in excess, it may be argued that these organelles play a modest role in fatty-acid synthesis.

This outline, and it is no more than an outline, of the main reaction pathways in mitochondria underlines the central importance of these organelles in intermediary metabolism.

4.7 The electron transport chain

During the TCA cycle, reducing equivalents are removed from oxidizable substrates and transferred to the coenzyme nicotinamide adenine dinucleotide (NAD^+) from which they pass to oxygen via a sequence of hydrogen and electron carriers in the mitochondrial inner membrane. The main components of the electron transport chain are shown below (figure 4.11).

NADH dehydrogenase is a large flavoprotein containing flavin mononucleotide (FMN) which oxidizes NADH by accepting two

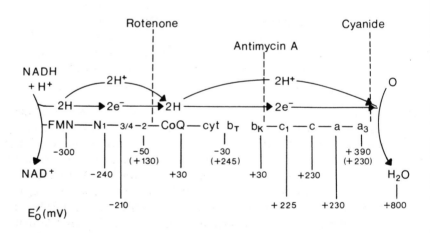

Figure 4.11 The mitochondrial respiratory chain showing the standard oxido-reduction potentials (E_0') of the carriers and the sites of action of the respiratory inhibitors rotenone, antimycin A and cyanide.
FMN: flavin mononucleotide. FMN and nonhaem irons N1, 2, 3 and 4 (or more) are the carriers in NADH dehydrogenase, followed by coenzyme Q (ubiquinone) and the cytochromes. The E_0' values of N2, cyt b_T and cyt a_3 can "slide" to the values in parenthesis under certain conditions (see Chance, 1973).

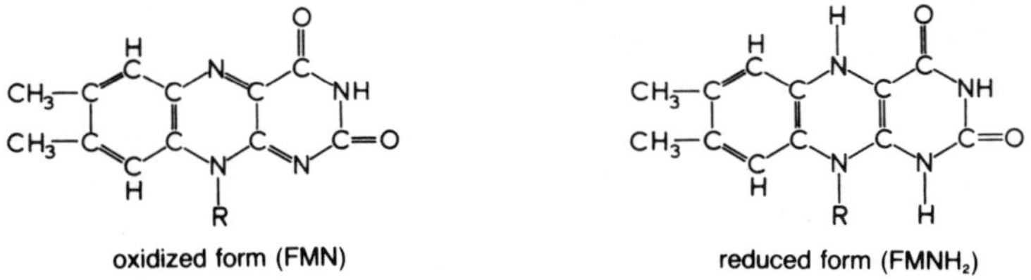

Figure 4.12   Reduction of the isoalloxazine ring of flavin mononucleotide (FMN). The R group is ribitol phosphate. FMN is tightly bound to the protein of NADH dehydrogenase.

hydrogens into its isoalloxazine ring (figure 4.12). Low-temperature electron paramagnetic-resonance (EPR) studies have shown that the enzyme contains at least four non-haem iron centres. These centres each contain 2–4 iron atoms covalently bound to protein by cysteine sulphur atoms, the centres being called N1, N2, N3 and N4 on the basis of the temperature at which their characteristic EPR spectra appear. In fact, electrons from $FMNH_2$ reduce the centres in the sequence N1-N3/N4-N2, each centre acting as a single electron carrier. The next carrier ubiquinone or coenzyme Q is a substituted benzoquinone thought to act in the lipid phase of the membrane as a shuttle between NADH dehydrogenase and succinic dehydrogenase and the cytochromes (figure 4.13).

The cytochromes are electron-transferring proteins with a haem prosthetic group—a porphyrin ring in which the four pyrrole nitrogens are co-ordinated to an iron atom that undergoes valency changes on oxido-reduction. They have characteristic spectral properties which are related to the haem environment and its side groups. In cytochrome c_1 and c the haem is covalently bound to the protein by cysteine side-chains and by co-ordination bonds between the iron and the side-chains of methionine and

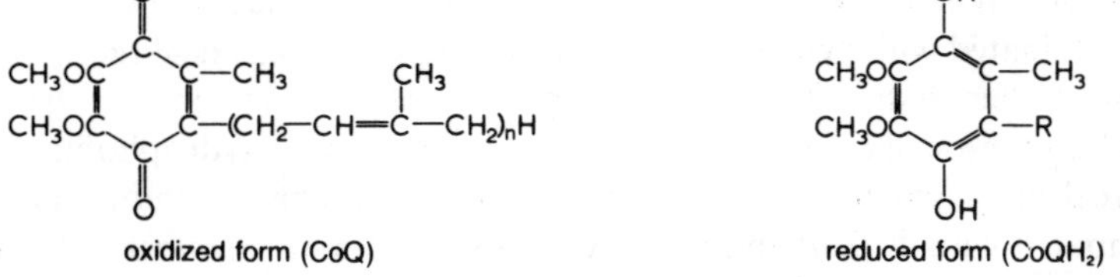

Figure 4.13   Oxidized and reduced forms of ubiquinone (coenzyme Q). In mammalian mitochondria $n = 10$. Note the lipid soluble nature of the molecule.

histidine residues of the protein (figure 4.14). Cytochrome classes a and b have different haem side-groups and are not covalently bonded. Mitochondria are now considered to have at least two cytochrome b species in the respiratory chain, cytochrome b_K (K signifying the original cytochrome b discovered by Keilin) and cytochrome b_T (T standing for transducer, because it was thought to be directly involved in energy transduction to the ATP synthesizing system). Cytochromes b_K and b_T are spectrally different, but they have not been physically separated, and it is

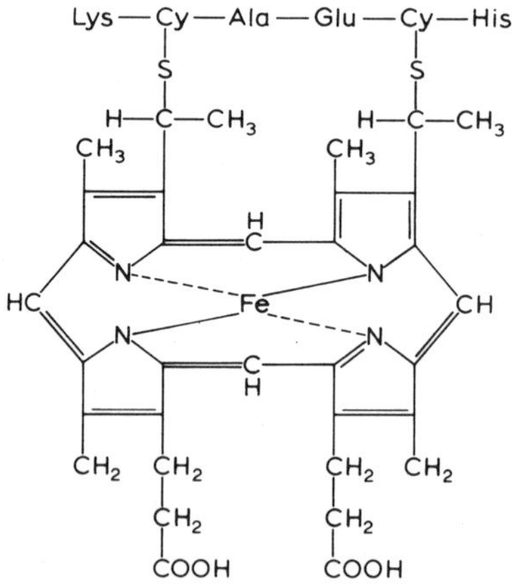

Figure 4.14   The haem group of cytochrome c. It is linked covalently to the peptide by the cysteine residues. The fifth and sixth co-ordination positions of the iron are occupied by the side-chains of a histidine and a methionine.

possible that they are structurally similar but in different membrane positions. When cytochrome b is extracted from the membrane, it is often contaminated with cytochrome c_1, suggesting a close physical relationship between them *in situ*. The haem and the absorption spectrum of c_1 are very similar to those of cytochrome c but the protein is more hydrophobic and is more deeply sited in the membrane. Cytochrome c is smaller than c_1 (12 000 compared with 40 000) and is easily removed from the membrane by high salt concentrations. Cytochrome c donates electrons to cytochrome oxidase (a-a_3), a large protein (200 000 MW) of seven subunits with two haems and two copper atoms. Cytochromes a and a_3 are spectrally distinct, although like cytochromes b_K and b_T they have not been physically separated. The a_3 moiety reacts directly with carbon monoxide, cyanide and oxygen, while a is the oxidant of reduced cytochrome c. The mechanism of electron transport through cytochrome oxidase is not fully understood, particularly the electron-carrying role of the copper atoms.

Plant mitochondria and those of lower eukaryotes frequently have more complex chains than typical animal mitochondria. Many plants have an extra NADH dehydrogenase and a cyanide-resistant cytochrome oxidase that together provide an alternative pathway for NADH oxidation that is not linked to ATP synthesis. The TCA cycle is also often more versatile than in animals, and can operate without pyruvate from glycolysis. This gives a high degree of flexibility to meet the different situations in organisms that have photosynthetic as well as respiratory systems. Palmer (1979) has usefully reviewed this subject.

The order of the carriers in figure 4.11 is based on kinetic experiments and reconstitution studies. Changes in oxido-reduction states of the carriers can be measured in intact mitochondria by dual beam spectrophotometry. This technique has been used in experiments with inhibitors that bind to specific carriers and prevent electron transport to the carriers on the oxygen side of the block, and in studies with electron donors and acceptors that interact at different points in the chain. The results of these and other approaches are consistent with the postulated order of carriers. Functional chains have been reconstituted from isolated components, section by section, and it has been clearly shown that NADH dehydrogenase, coenzyme Q and cytochromes b, c_1, c and a-a_3 are all necessary. As indicated in figure 4.11, with one or two minor exceptions, the standard redox potential (E_0') of each carrier in the model is more positive than that of the preceding carrier, a feature that is consistent with electron flow through the chain from NADH to oxygen. In fact the situation in the CoQ to cytochrome c region is not at all clear, and linear-flow models like the one depicted are almost certainly gross simplifications of electron transport, a point we shall return to.

ATP synthesis linked to electron transport

In functional mitochondria, electron transport is coupled to the synthesis of ATP from ADP and inorganic phosphate. This is illustrated by the experiment in figure 4.15. Mitochondria incubated in a physiological buffer show a very low rate of electron transport (oxygen reduction) even when

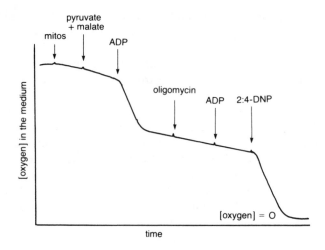

Figure 4.15 Respiratory control and uncoupling of oxidative phosphorylation in mitochondria. Details in text.

oxidizable NAD^+-linked substrates are present. However, on adding ADP, the respiration rate increases significantly and remains high until all the added ADP has been phosphorylated to ATP, a point that can be checked by assaying the ATP produced. This control of respiration rate by ADP is explicable if electron transport is coupled to phosphorylation, for the essence of a coupled reaction is that inhibition of one half of the couple (in this case inhibition of phosphorylation by lack of ADP) prevents the other half (electron transport) from taking place.

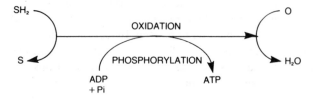

Further evidence of coupling is shown in that ADP does not stimulate respiration if the ATP synthetase enzyme is inhibited by the antibiotic oligomycin. Under conditions where ATP synthesis is prevented by lack of ADP or Pi, or by inhibition of ATP synthetase, the maximum rate of respiration can be induced only by uncoupling electron transport from ATP synthesis. This can be done by a range of chemical reagants, including phenols, detergents and certain peptides. When mitochondria are uncoupled, the free energy released during electron transport is dissipated as heat rather than used to synthesize ATP.

Data from experiments like that shown in figure 4.15 and from direct measurements of ATP have indicated that 3 molecules of ATP are synthesized for every atom of oxygen reduced by electrons from NADH, i.e. an ATP:O ratio of 3. In the case of substrates that feed electrons into the respiratory chain at Q (succinate) and at cytochrome c_1/c (e.g. ferrocyanide) the ATP:O ratios have been calculated as 2 and 1 respectively. This has led to the view that there are three spans where energy transfer occurs between the respiratory chain and the ATP synthesizing system (figure 4.16). It has proved extremely difficult to define these coupling sites. In figure 4.16 the phosphorylations are shown linked to the chain at the three spans where large decreases in free energy occur during electron transport, but there is no good evidence for direct physical links between electron carriers and the phosphorylation processes. Irrespective of the nature of the sites or the mechanism involved, oxidative phosphorylation is highly efficient. Calculations based on information about the steady-state concentrations of ADP, Pi and ATP and the oxido-reduction states of the carriers indicate that the synthesis of 3 molecules of ATP would require almost 100% of the free-energy change for 2 electrons moving between NADH and oxygen. Electron transport can be reversed through the first two coupling sites if the mitochondria are exposed to very high [ATP]/[ADP][Pi]ratios, the energy coming from ATP breakdown. One ATP can drive two electrons back through one "site"; this stoichiometric reversal and other energetic considerations confirm that energy transfer in oxidative phosphorylation is efficient with little loss.

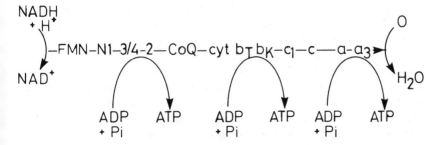

Figure 4.16 Energy transfer "sites" between electron transport and ATP synthesis.

4.8 Models of oxidative phosphorylation

Early hypotheses on energy coupling proposed that the energy was tapped from the respiratory chain through a series of chemical intermediates, the last being a phosphorylated intermediate that donated its phosphate to ADP in a reaction catalyzed by ATP synthetase. Uncouplers were thought to act by inducing hydrolysis of one or other of the intermediates, thus preventing energy transfer to the ATP synthetase. However no high-energy chemical intermediates have been isolated to date, and it is difficult to envisage a structure that would be susceptible to the wide range of chemically different uncoupling agents. Consequently, although the hypothesis has theoretical appeal, it lacks experimental support.

In 1961 a chemiosmotic hypothesis was proposed by Peter Mitchell. This postulated that the electron transport chain was arranged vectorially in the mitochondrial inner membrane so that the transport of reducing equivalents along the chain created a transmembrane electrochemical potential difference of protons ($\Delta\mu_{H^+}$) which drove the membrane-sited ATP synthetase. This hypothesis has been mentioned in the context of photophosphorylation (chapter 3) and the principles are essentially the same. In both cases electron transport pumps H^+ ions across the membrane, and the resulting $\Delta\mu_{H^+}$ drives H^+ back through a proton-translocating ATP synthetase. This can work only if the membrane has a low natural permeability to H^+, since non-specific back leakage of protons would, of course, prevent formation of a large enough $\Delta\mu_{H^+}$ to drive ATP synthesis. The model proposes that uncouplers act by damaging the membrane, or conducting H^+ through, thus collapsing the vital $\Delta\mu_{H^+}$ and removing the proton motive force necessary to drive ATP synthesis.

The chemiosmotic model has been tested and developed by Mitchell who received the 1978 Nobel Prize for Chemistry for his work. A key experiment showing that proton translocation is indeed linked to electron transport is shown in figure 4.17. Mitochondria were incubated under anaerobic conditions, with the ATP synthetase inhibited by oligomycin and an NAD^+-linked substrate present. Introduction of a pulse of oxygen in saline allowed electron transport to start, and this was associated with the appearance of H^+ outside the mitochondria, as recorded by a glass pH electrode. This would quickly lead to a transmembrane $\Delta\mu_{H^+}$ that would oppose further H^+ ejection and so, in such experiments, precautions must be taken to ensure that a limiting $\Delta\mu_{H^+}$ is not reached. When the added oxygen was all reduced and electron transport had ceased, the protons

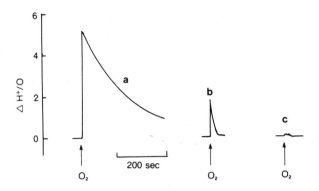

Figure 4.17 Net proton translocation by rat liver mitochondria on injecting a small amount of air-saturated KCl into an anaerobic suspension containing the NADH linked substrate β-hydroxybutyrate. The effects of the uncoupler, 2:4-dinitrophenol and the lipid solvent, Triton X 100 are shown in traces *b* and *c* respectively. This is characteristic of treatments that damage membranes or increase their permeability to protons. Treatments that prevent the development of a transmembrane protonic difference also prevent ATP synthesis. The decay of the proton pulse in trace *a* is due to the equilibration of the ΔpH established across the membrane during the brief period of electron transport (experimental details in P. Mitchell & J. Moyle *Nature* **208**, 142–151, 1965).

slowly equilibrated back across the membrane, a process that was speeded up by addition of uncoupler. The net number of protons translocated out in association with the transport of two electrons to oxygen through different segments of the chain was computed from experiments using different electron donors and acceptors. This and other data have been interpreted by Mitchell in terms of the model in figure 4.18. In this model the FMN of NADH dehydrogenase is considered to translocate H atoms from the matrix side (M) to the cytosol side (C) of the membrane where the electrons are removed by the non-haem irons and H^+ is released to the medium. The electrons return to the M side where, with H^+ from the matrix, they reduce the semiquinone QH. The resulting QH_2 is oxidized to QH by cytochrome b_T on the C side with H^+ release to the medium. QH is then further oxidized to Q by cytochrome c_1 with subsequent H^+ release. The Q is then reduced on the M side by electrons from cytochrome b_K and H^+ from the matrix, and so the cycle continues. The electrons from cytochrome c_1 pass across the membrane through cytochrome oxidase and reduce oxygen.

Turning to ATP synthetase, mitochondria in which electron transport has been inhibited by rotenone and Antimycin A hydrolyze added ATP with a stoichiometry of $2H^+$ ejected per ATP hydrolyzed. Oligomycin stops both ATP hydrolysis *and* proton ejection, showing that ATP synthetase is the key factor. Assuming that the reaction is reversible, ATP synthesis would be accompanied by a net uptake of 2 H^+ as depicted in figure 4.18. Under steady-state phosphorylating conditions, protons would be translocated out by the respiratory chain at the same rate as they were driven in through the ATP synthetase and other routes. The model in figure 4.18 indicates that 6 H^+ ions are driven out during electron flow from NADH to oxygen. The $\Delta\mu_{H^+}$-driven return flow of these through the ATP synthetase would give $6H^+/2H^+$ or 3 ATP molecules—the classically

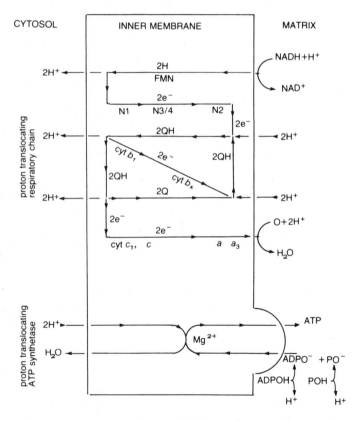

Figure 4.18 A representation of the possible arrangements of the carriers and ATP synthetase in the inner membrane of the mitochondrion to account for the observed stoichiometries of proton translocation (after Mitchell).

accepted number for the NADH to oxygen span. However, this neat scheme is complicated by evidence that the transport of ADP and Pi into the matrix and ATP out must be balanced by the net uptake of 1 H^+ (figure 4.21), meaning that only 2 ATPs could be synthesized from the $6H^+$ ($2 \times H^+$ for transport and $2 \times 2H^+$ for phosphorylation). Hinkle and Wu (1979) argued that most published data, in fact, indicate an ATP:O of 2 for NADH-linked substrates, and that the generally accepted figure of 3 may be wrong. However, the counter-argument is that much of the published data may have been derived from preparations with a proportion of uncoupled mitochondria. In the meantime Mitchell and his coworker Jennifer Moyle incline to the view that Pi and nucleotide transport might occur through a Ca^{2+} recycling system without the net translocation of $1H^+$ (Mitchell, 1979).

Although the details of oxidative phosphorylation are still obscure, the broad lines of the chemiosmotic hypothesis are supported by various findings; for example, a substantial $\Delta\mu_{H^+}$ of 230 mV or more exists across the inner membrane. This compares favourably with the theoretical $\Delta\mu_{H^+}$ required for ATP synthesis calculated by Mitchell, although some workers think it should be greater. In passing, $\Delta\mu_{H^+}$ includes both a chemical concentration difference of H^+ (ΔpH) and an electrical difference $\Delta\psi$. ($\Delta\mu_{H^+} = \Delta\psi - RT/nF \Delta$pH). In mitochondria, the $\Delta\psi$ component is greatest and ΔpH is relatively small; in chloroplasts, the converse holds. Another difference between these organelles is that in mitochondria the potential is positive *outside*; in chloroplast thylakoids the potential is positive *inside*.

As in chloroplasts, ATP synthesis can be induced in mitochondria in the absence of electron transport (i.e. in the presence of respiratory inhibitors) by imposing artificial pH and $\Delta\psi$ differences across the membrane (figure 4.19). Other evidence has been afforded by reconstituted systems. Racker and coworkers incorporated purified cytochrome c and cytochrome oxidase into phospholipid vesicles, and showed that electron transport generated a $\Delta\mu_{H^+}$ that was capable of driving ATP synthesis when the ATP synthetase was also incorporated into the membrane. This, experiment showed that the third segment of the respiratory chain could be reconstituted and could operate on chemiosmotic principles. There have been various elegant experiments of this nature done in the laboratories of Racker and Skulachev.

Studies of the molecular architecture of the inner membrane are still at a relatively early stage. The membrane is about 75% protein and 25% lipid, and is evidently a three-dimensional array of proteins in a lipid matrix.

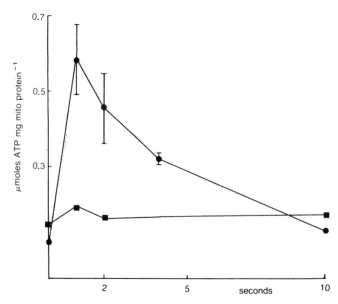

Figure 4.19a ATP synthesis driven by ΔpH
The pulse of ATP synthesis in figure 4.19*a* was obtained on exposing respiratory-inhibited mitochondria to a transitory ΔpH of 4.6 units (acid outside). In this case $\Delta\psi = 0$ since valinomycin, a K^+ ionophore, was present allowing charge equilibration across the membrane. As the imposed ΔpH decayed, the ATP synthetase reversed, causing ATP hydrolysis. No ATP synthesis occurred in presence of 2:4-DNP, oligomycin or at significantly smaller ΔpH (lower trace ■) (from Reid, Moyle & Mitchell, 1966).

Approximately 30–40% of the membrane protein is electron carriers and ATP synthetase. Negative staining shows spheres of about 10 nm on the M side. These have been identified as the F_1 parts of the ATP synthetase and probably reside rather deeper in the membrane than negative staining suggests. The catalytic sites for ATP synthesis are on F_1 and the membrane factor of the enzyme appears to constitute the proton channel and the site of oligomycin inhibition. It is now established that cytochromes c_1 and c are on the C face of the membrane, and that cytochrome oxidase extends across the membrane with its subunits possibly arranged as shown in figure 4.20. The oxygen reducing site (a_3) is on the M face and the cytochrome c oxidizing site (a) is near the C face. Cytochrome b is deeply embedded in the membrane as it cannot be labelled from either side. The substrate acceptor sites of the NADH and succinic dehydrogenases are on the M side, but little is known about the orientation of the remainder of these molecules. The available information therefore points to an anisotropic arrangement of

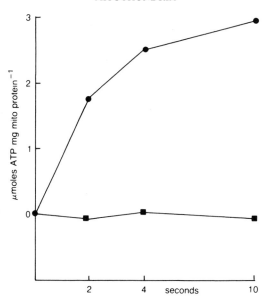

Figure 4.19b   ATP synthesis in respiratory-inhibited submitochondrial particles (vesicles) on artificially establishing an inwardly directed K^+ gradient ($\Delta\psi$) which tends to drive H^+ *out* through the ATP synthetase. Since smp are inside out relative to intact mitochondria, the polarity of proton-driven phosphorylation is opposite to that in mitochondria. No ATP synthesis occurred in presence of inhibitors of ATP synthetase (lower trace) (Broughall, Reid & Wright in Reid (1974)).

the respiratory chain. With the exception of coenzyme Q, the carriers all seem to be in relatively fixed positions, so that electron and hydrogen transfer would occur by small rotational and translational diffusions rather than large transmembrane movements. Rapid progress should occur in the next few years using cross-binding reagents, EPR and physical techniques. The obvious questions are whether the ATP synthetase is in physical contact with specific electron carriers, how the dehydrogenases are positioned relative to the cytochrome chain (there appears to be only 1 NADH dehydrogenase for every 5 cytochrome chains) and, of course, the actual mechanism of proton translocation.

Oxidative phosphorylation has been an exciting and controversial research area for many years, and is likely to continue so for a long time. The stoichiometry of H^+ translocation is by no means settled. It has been claimed by Lehninger and Brand that $9-12H^+$ are translocated out during 2-electron transport from NADH to oxygen in contrast to the $6H^+$ found by Mitchell. Mitchell disputes these claims, and the reader is referred to

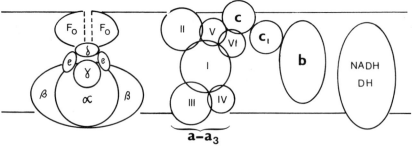

Figure 4.20   A model of the locations of the electron carriers and ATP synthetase in the mitochondrial membrane, based on experiments on the responses of the proteins to cross-linking reagents and antibodies, carried out in several laboratories. There is no evidence for physical contact between NADH dehydrogenase and cytochrome b, but Q is known to act as a mobile hydrogen-carrying shuttle. Also there is no evidence for direct physical contact between ATP synthetase and respiratory carriers. There are uncertainties over the structure of cytochrome oxidase and the ATP synthetase. The structure shown for the latter is based on an $\alpha_2\beta_2\gamma_2\delta l_2$ stoichiometry for F_1. It is thought that the membrane factor subunits are arranged so that a proton channel runs through to F_1.

papers by Brand (1979) and Mitchell and Moyle (1979) for information on this controversy. If the high stoichiometry is correct, the problem of finding H^+ ions for Pi, ADP and ATP transport (page 111) will be solved, but models of the chain (such as the one in figure 4.18) will need to be substantially revised, possibly in the direction of models that explain H^+ translocation in terms of Bohr effects. Some researchers think that the electron carriers and associated proteins may act as a series of Bohr domains in which electron transfer causes structural changes, e.g. altered bond length that changes the dissociation constant of an ionizable group so that protonation/deprotonation occurs, the group gaining a proton on one side of the membrane and losing it to the other. Recently it has been claimed by Wikstrom, Chappell and others that cytochrome oxidase translocates protons out through the membrane during electron transport. If this is confirmed, it will strengthen the idea that Bohr pumps are involved in respiratory chain H^+ translocation.

Another controversial area in bioenergetics is the mechanism of the H^+-translocating ATP synthetase. There is no consensus on how this might operate, and many fundamental questions remain unanswered. Finally, as shown in figure 4.11, some electron carriers, namely N2, cytochrome b_K and cytochrome a_3, have E_0' values that change or "slide" under different physiological conditions. The significance of this is far from understood, but students who wish to pursue it will find a speculative lead in Chance (1972).

4.9 Anion and cation carriers in the mitochondrial membrane

The foregoing discussion has underlined the importance of the mitochondrial inner membrane as a device for separating electric and osmotic potential, and attention has been drawn to its low natural permeability to protons. The membrane must also permit entry of solutes for metabolism and biosynthesis, and the exit of at least some of the products. This is achieved through some 12 different exchange-diffusion carriers for anions in the inner membrane. Of these, the carriers for pyruvate, Pi, ADP and ATP are present in all mitochondria, which is not surprising since all mitochondria oxidize pyruvate and synthesize ATP from ADP and Pi. As figure 4.21 shows, Pi, pyruvate and glutamate all enter by their respective translocases in exchange for OH^- ions, and it is thought that the entry and accumulation of these solutes is directly influenced by the ΔpH component of the transmembrane $\Delta\mu_{H^+}$ set up by the respiratory chain. A large ΔpH would favour accumulation of the solutes, as it would provide a downhill gradient for the exit of OH^-. The Pi translocase plays a key role in transport processes, because it results in Pi being concentrated in the matrix, and this provides a driving force for the uptake of dicarboxylic acids like succinate and malate, since the internal Pi tends to move down its concentration gradient and out of the mitochondrion on the dicarboxylic acid translocase in exchange for a dicarboxylic acid. Some mitochondria have a tricarboxylic acid translocase for citrate and its analogues, but even this is indirectly linked to the Pi translocase because malate which enters in exchange for internal Pi can recycle out on the tricarboxylic acid translocase in exchange for citrate. The scheme in figure 4.21 should not be taken too rigidly, for the nature of the carriers permits great flexibility. The dicarboxylic acid carrier allows exchange of dicarboxylic acids, and the tricarboxylic acid translocase does likewise for tricarboxylic acids.

The presence of carriers that permit the entry and exchange of substrate intermediates like succinate, malate, citrate and α-ketoglutarate dispels the impression given by some metabolic charts that the TCA cycle is "locked up" in the mitochondrion with only one entry point from the cytosol pyruvate. A wide range of substrate anion carriers is understandable in liver which receives many different metabolic intermediates from the

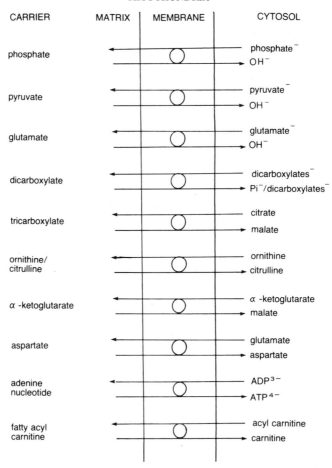

Figure 4.21   Carriers present in the inner membrane of rat liver mitochondria.

intestine. It is important that these can enter the mitochondrion for oxidation or conversion to phosphoenolpyruvate for glycogen synthesis. In contrast to liver mitochondria, those in blowfly flight muscle have only a carrier for pyruvate, in addition to the ubiquitous phosphate and ADP/ATP carriers. Pyruvate and α-glycerophosphate are the main substrates, and a battery of other carriers would provide leak points for TCA intermediates, undesirable in view of the very high respiratory rate demanded of these mitochondria. The α-glycerophosphate never enters the matrix, being oxidized from the outside by a membrane-bound dehydrogenase which passes its reducing equivalents to CoQ and the respiratory chain.

The adenine nucleotide translocase operates a reversible 1:1 exchange of ADP/ATP, ADP/ADP and ATP/ATP. However, on the cytosol side, it preferentially binds ADP and on the matrix side, ATP. Therefore typically it translocates ADP in and ATP out. Considering the charge difference between ADP and ATP, the translocase would seem to operate electrogenically, the movement of ADP in and ATP out being assisted by the $\Delta\psi$ component of the transmembrane $\Delta\mu_{H^+}$.

Mitochondria can actively accumulate certain cations, particularly Ca^{2+}. The Ca^{2+} translocator has a very high affinity for the cation, and the energy for transport is provided by electron transport, probably through the intermediacy of the transmembrane $\Delta\mu_{H^+}$. Cytosol Ca^{2+} concentrations are important moderators of many cell activities, and it seems likely that the uptake and release of Ca^{2+} plays a regulatory role in cell physiology.

4.10 The outer membrane

The outer membrane is composed of some 50% protein and 50% lipid, and contains several enzyme activities, namely monoamine oxidase, NADH-cytochrome b_5 reductase, acyl-CoA synthetases, and nucleoside diphosphate kinase. It has attracted less attention than the inner membrane, and its functions are poorly understood. It is relatively impermeable to molecules larger than 12 000 MW, and consequently may screen the important bioenergetic reactions in the inner membrane from incompatible proteins. It may also protect the inner membrane against osmotic rupture; mitochondria are often seen in a swollen state, with the inner membrane hard against the outer. It has been suggested that it is a modified extension of the endoplasmic reticulum (ER) (its NADH-cytochrome b_5 reductase is immunologically identical to the ER enzyme) and may serve as a transport channel for proteins and lipids for mitochondrial biosynthesis; however, there is little evidence for this function. The outer membrane creates an intramembrane compartment about which little is known, apart from the fact that it contains considerable amounts of adenylate kinase, which catalyzes the reaction

$$AMP + ATP \rightleftharpoons 2\ ADP$$

Adenylate kinase permits AMP, the end product of many reactions, to be returned to the main phosphorylation cycle in the form of ADP. AMP inhibits the adenine nucleotide translocator in the inner membrane, and it is thought that one advantage of the adenylate kinase being present near the translocator is that it would maintain the local concentration of AMP low enough not to affect the translocator. The monoamine oxidase enzymes on the outer membrane oxidatively deaminate dietary amines and certain neurotransmitters. Recently we have shown that MAO activity towards some but not all substrates is influenced by the bioenergetic state of the mitochondria (Smith and Reid, 1978), particularly the oxido-reduction state of the electron carriers. It is not clear how MAO activity in the outer membrane can be affected by events in the inner membrane. However, it has been shown in Mayer's laboratory that much of the MAO activity towards tyramine is located on the inner surface of the outer membrane. So it is tempting to think that the activity of MAO may be influenced by the inner membrane, through the physical points of attachment frequently observed between the two membranes. There are aspects of cell physiology outside the scope of this book where regulation of MAO activity by the redox state of the mitochondria would make sense.

CHAPTER FIVE

MICROBODIES

UNLIKE CHLOROPLASTS AND MITOCHONDRIA WHOSE EXISTENCE WAS KNOWN to the light microscopists long before their biochemistry was investigated, the cellular enzymic activities associated with microbodies were assayed before these particles were recognized as a distinct class of subcellular organelle. The name "microbody" was originally used to describe a particle in mammalian liver cells, circular in profile and $0.2–1.5$ μm in diameter, with a sparsely granular matrix, and bounded by a single membrane. These animal-cell microbodies were subsequently shown to contain flavin-dependent oxidases and catalase, and to be centres of non-phosphorylating oxidations, notably the oxidative deamination of certain amino acids:

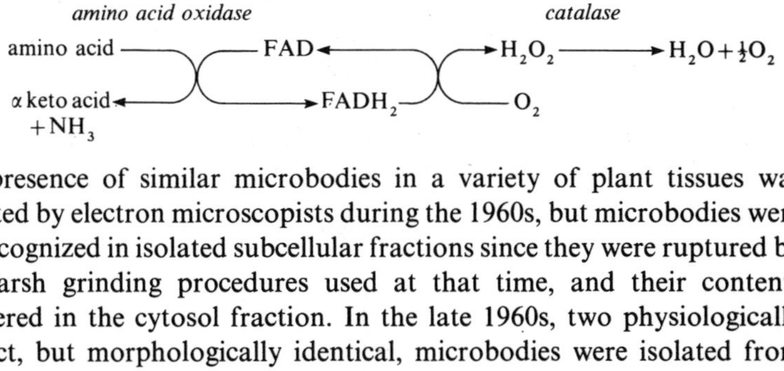

The presence of similar microbodies in a variety of plant tissues was reported by electron microscopists during the 1960s, but microbodies were not recognized in isolated subcellular fractions since they were ruptured by the harsh grinding procedures used at that time, and their contents recovered in the cytosol fraction. In the late 1960s, two physiologically distinct, but morphologically identical, microbodies were isolated from plant tissues (using gentle chopping procedures) and shown to be of major importance in plant cell metabolism. Microbodies can be separated from mitochondria on discontinuous sucrose density gradients when they lose water through their permeable membrane and behave as if they have a specific density of $1.24–1.26$. This specific density is sufficiently different from those of mitochondria, chloroplast fragments and cellular membranes, to allow separation of the microbodies from other cellular material. In these enriched suspensions of microbodies, their appearance resembles the appearance of microbodies in the tissue cells from which the suspensions were derived.

The two types of plant microbody are more sophisticated than animal microbodies and known as *peroxisomes* and *glyoxysomes* to distinguish their different metabolic roles. Peroxisomes are found in leaf cells in close association with the chloroplasts, and are the site of glycolate oxidation. Glyoxysomes are particularly abundant in cells of germinating fatty seeds, and contain all the enzymes required for the conversion of fatty acids to succinate. Since they are also the site of monoglyceride hydrolysis, the glyoxysomes are key organelles in the gluconeogenic synthesis of sucrose from triglyceride in the utilization of the fat of the endosperm of germinating seeds. In common with animal cell microbodies, both peroxisomes and glyoxysomes contain flavin-dependent oxidases and catalase, and are able to oxidize endogeneous substrates with intermediate formation of hydrogen peroxide, subsequently broken down by catalase to oxygen and water, or by catalase and other peroxidases to oxidize other substrates.

5.1 Peroxisomes

Leaf peroxisomes were first isolated from spinach leaf homogenates by Tolbert's group in Michigan in 1968. Since that time, peroxisomes have been isolated from the leaves of several different species of C3 and C4 plants, and their presence recognized in electron micrographs of mesophyll cells of many leaves.

The first suspensions of peroxisomes isolated by Tolbert were shown to contain the bulk of the catalase, glycolate oxidase and hydroxypyruvate

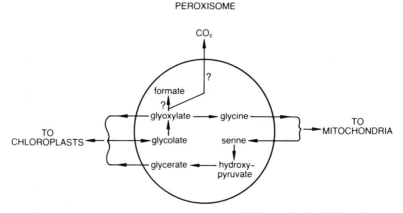

Figure 5.1   The reactions of the glycolate pathway which occur in the leaf peroxisome.

reductase of the leaf cell, and it was suggested they were implicated in glycolate metabolism. The formation of phosphoglycolate from RubisP in photosynthesizing chloroplasts under conditions of high partial pressures of oxygen is discussed in chapter 3. Phosphoglycolate is converted to glycolate in the chloroplast from which it is released into the cytosol. Its further metabolism takes place in the peroxisome whose enzyme spectrum is illustrated in figure 5.1. The two aminotransferases catalyzing the formation of glycine from glyoxylate, and of hydroxypyruvate from serine, have also been shown to be localized in leaf peroxisomes, so serine hydroxymethyltransferase is the only enzyme of the glycolate (figure 5.2) pathway which has not been shown to be present in the peroxisome. The conversion of glycine to serine appears to be localized in the mitochondria of leaf cells. Present evidence therefore suggests that peroxisomes, chloroplasts and mitochondria may all have the capacity to decarboxylate products of the glycolytic pathway, and this example of intracellular cooperation will be discussed in chapter 8.

A further example is afforded in developing seeds where the complete reactions of glyconeogenesis involves 3 organelles—spherosomes, in which triglycerides are hydrolyzed to glycerol and fatty acids; glyoxysomes in which fatty acylCoA is converted to succinate, which is tranformed in the mitochondria to oxaloacetate, whose subsequent conversion to sucrose occurs in the cytosol.

5.2 Glyoxysomes

In 1969 Beevers discovered microbodies in the endosperm cells of germinating castor beans, demonstrated their functional role in the glyoxylate cycle, and named them glyoxysomes. The successful isolation of glyoxysomes became possible after gentler methods of chopping tissue with razor blades were introduced, and the separation of glyoxysomes from

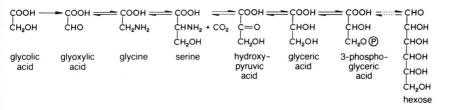

Figure 5.2 The glycolate pathway.

mitochondria was achieved on sucrose density gradients where the density of the glyoxysomes ($1 \cdot 24$–$1 \cdot 26$ g cm^{-3}) allows their separation from mitochondria (density $1 \cdot 16$–$1 \cdot 19$ g cm^{-3}). The role of the glyoxysomes in the conversion of fat to sucrose is not, however, restricted to the crucial stages of the glyoxylate cycle; the acetyl CoA consumed in this cycle is also synthesized in the glyoxysome from fatty acids. Long-chain fatty-acyl CoA derivatives are converted to acetyl CoA with concomitant uptake of oxygen and formation of NADH. Thus the entire sequence from fatty-acid activation to succinate formation occurs within the glyoxysome (figure 5.3).

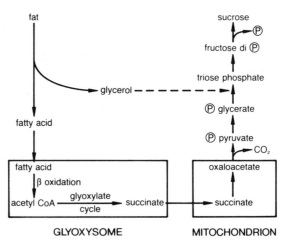

Figure 5.3 The pathway for the conversion of fat to sucrose in the organelles and cytosol of the endosperm of castor bean seedlings during early germination (after Beevers in *Recent Advances in the Chemistry and Biochemistry of Lipids*, Academic Press, pp. 287–299 (1975)).

5.3 The origin and development of microbodies

Catalase is the first enzyme to be detected in microbodies during germination in the castor bean endosperm, and malate synthetase and isocitrate lyase are detectable almost immediately afterwards. The activity of the latter two enzymes rises dramatically; it reaches a peak of activity five days after germination, and then declines. The enzymes are synthesized *de novo* and assembled in the glyoxysome which appears to develop directly by budding from the endoplasmic reticulum. Connections between developing microbodies and the cisternae of the endoplasmic reticulum have been demonstrated in electron micrographs, and developing glyoxysomes have been shown to contain ^{14}C-choline-labelled phosphatidyl choline derived from the endoplasmic reticulum. Increases in the number of microbodies per cell have been shown in castor bean endosperm during germination, when the final number of glyoxysomes is double the number of mitochondria per cell. There are also some indications that glyoxysomes may be converted into peroxisomes during greening of the seed leaves (cotyledons) of plants such as cucumber. Since the two types of microbody are morphologically indistinguishable, evidence for the change from glyoxysome to peroxisome is based on changes in the enzymic spectrum of the organelle. In the germinating seeds of cucurbits, the cotyledons show high activity of the glyoxysomal enzymes, malate synthetase and isocitrate lyase, which decrease when the cotyledons emerge from the seed, become exposed to light and begin to green. Concomitantly the enzymes of glycolate metabolism characteristic of peroxisomes increase. In water melon cotyledons, manipulation of the light conditions causes an increase in the peroxisomal enzymes without loss of the glyoxysomal ones, suggesting that peroxisomes may be formed in conditions where glyoxysomes are still preserved. It is not known whether the two microbodies are interconvertible.

CHAPTER SIX

ENDOPLASMIC RETICULUM, GOLGI COMPLEXES AND SECRETORY VESICLES

6.1 Protein secretion

During the 1950s an avalanche of electron microscopic (E.M.) studies showed that the cytoplasm of most plant and animal cells contained a complex membrane system of canaliculi and cisternae. This was named the endoplasmic reticulum (ER) by Palade and was divided into rough ER, characterized by ribosomes on the outer surfaces of the cisternae, and smooth ER, which was devoid of ribosomes (figure 6.1). The rough ER was 40–50 nm wide, approximately half the width of the smooth ER elements. A third membrane system, known as the Golgi body or apparatus, since it was first reported by Golgi in 1898, was confirmed by E.M. and shown to be a relatively discrete interconnected fretwork of flattened sacs (saccules), tubules and vesicles that could fairly be termed an organelle. The finding that ER and Golgi bodies were particularly well developed in synthetically active cells like exocrine pancreas, liver, collagen-secreting fibroblasts and polymer-secreting plant cells, suggested that they were involved in synthetic and secretory activity. Subsequent investigations brilliantly vindicated these predictions.

One of the first steps that led to the discovery that ER was a centre for protein synthesis occurred in 1940, when Claude separated from liver homogenate a granular RNA-rich fraction sedimenting at higher g forces than the mitochondrial fraction which he named the *microsome fraction*. Ten years later, in the post-war cell biology boom, it was shown that the microsome fraction could synthesize protein. Not only did it incorporate ^{14}C amino acids into proteins faster than other cell fractions, including nuclei and mitochondria, but the fraction always contained considerable amounts of the main protein synthesized by the tissue from which it had been isolated. By the end of another ten years, it was accepted that the microsome fraction was mainly vesicles formed from ER during the tissue homogenization, and that its protein-synthesizing activity reflected that of the rough ER. There was both morphological and biochemical evidence for

124

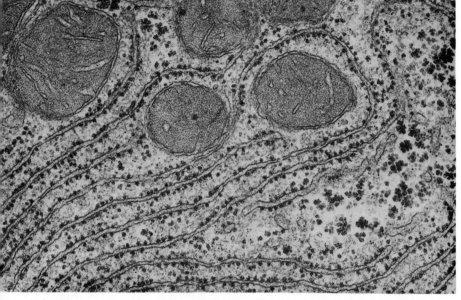

Figure 6.1a   Rough endoplasmic reticulum with attached ribosomes, and mitochondria in a parenchymal cell of rat liver. (Dr. Gillian Bullock) (× 40 000).

this. Morphologically, microsomes were membranous vesicles with ribosomes on their surfaces, very similar to rough ER. Palade showed that in cells which were damaged just prior to fixation with osmium, the ER became vesiculated into membrane-bound vesicles resembling microsomes. Furthermore 14-C amino acids administered to animals could be demonstrated radioautographically in the rough ER of tissue slices and recovered as polypeptides in the microsome fraction. Since protein synthesis could demonstrably occur *in vitro* on isolated ribosomes without associated membranes, and the ribosomes in bacteria were not attached to internal membranes, it was not immediately obvious why so many of the ribosomes in cells of higher organisms were bound to ER membranes. A clue was provided by the finding that, although the ribosomes were on the outside of microsomal vesicles, the proteins they synthesized were usually trapped in the vesicles rather than released to the exterior. This was corroborated by radiochemical studies on tissue slices by Palade, Siekevitz and others, which showed that, in cells that synthesized proteins for export, the growing polypeptide chains were apparently threaded directly from the ribosomes through the ER membrane into the cisternae (figure 1.4). The firmest evidence for this was a demonstration by Sabatini and Blobel (1970) that, when proteolytic enzymes were added to microsomes actively synthesizing protein, they cleaved the growing polypeptide chains into two parts—chains with their C′ ends still attached to the ribosomes, and other chains that could only be recovered from inside the microsomes. The simplest interpretation of this observation was that the proteins had been attacked by the enzyme at very short exposed regions between the ribosome and the ER membrane.

How are proteins for export segregated into the ER, whereas many other proteins required for internal use are not? There is nothing specific about the ribosomes that bind to rough ER, since they can be experimentally removed from microsomes and replaced by other ribosome populations. The binding sites, as visualized by freeze-fracture E.M. and their susceptibility to protease action, are intrinsic moieties that are quite deeply seated in the membrane. Recently Sabatini and co-workers have described

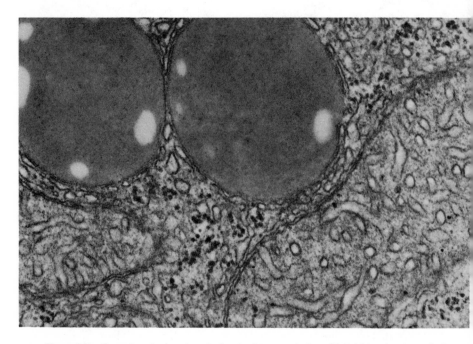

Figure 6.1b Smooth endoplasmic reticulum in close association with lipid droplets in a cell of the zona fasiculata of rat adrenal gland. (Dr. Gillian Bullock) (× 53 000).

two ribosome-binding proteins that are present in rough ER but not in smooth ER, and these may be constituents of the sites. Initially it was suggested that proteins for export had specialized N' terminal ends and that, as the end emerged from the ribosome, it provided a signal for a conformational change in the site proteins that led to ribosome binding and the opening up of an aqueous tunnel through which the growing (nascent) polypeptide passed into the ER lumen. All export proteins investigated, with the exception of ovalbumin, have indeed been found to have removable prepieces—N' terminal segments 15–25 amino acids long. However, the original "signal" or "threading" hypothesis seems incompatible with the fact that the prepieces are highly variable and show no common signal sequence; they are also too hydrophobic to be natural choices for leader sequences designed to penetrate aqueous tunnels (although characteristically the first amino acid is charged). Several alternatives have been proposed including a direct transfer model (figure 6.2) by Von Heijne and Blomberg (1979) that dispenses with tunnel proteins and aqueous pores. This proposes that the charged amino acid associates with the charged cytosol face of the ER, and that the hydrophobic prepiece that follows penetrates the lipophilic core of the membrane as it is synthesized, in a membrane-spanning α-helix (21 amino acids would be enough to span the membrane). The amino acids of the protein proper, many of them hydrophilic, would then follow, forming a loop; the binding of the ribosome to the membrane is envisaged as being strong enough to force the hydrophilic residues into the membrane. As the size of the nascent chain on the lumen side increased, it would assist the transfer by providing pulling power, particularly to bring the C' end through the membrane after synthesis has ceased. The membrane-bound

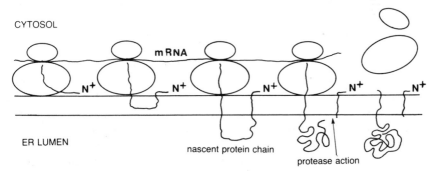

CYTOSOL

mRNA

N+   N+   N+   N+   N+

ER LUMEN

nascent protein chain

protease action

Figure 6.2 Diagrammatic representation of the synthesis of secretory proteins and their insertion through the membrane of the rough ER into the cisterna, by a direct transfer model.

prepiece would then be cleaved by a protease and the protein released into the lumen. At the time of writing, Blobel's group has reported that ovalbumin has a prepeptide-like signal equivalent, located more than 200 residues into the molecule, that is the basis for the ability of ovalbumin to compete with other presecretory proteins for entry into microsomes. It is not yet known how the very long N′ terminal sequence of over 200 residues that precedes the hydrophobic sequence is transferred into ER, but its existence is a further argument against signal/threading models. Most of the present data on protein transfer are consistent with loop or direct transfer mechanisms, but there are many unanswered questions on intramembrane interactions.

The transport route for export proteins has been deduced from experiments where metabolizing tissue slices of exocrine pancreas and intestine were given a pulse of ^{14}C amino acids and the sites of radioactivity detected at various times after the pulse. Proteins first appear in the rough ER and then proceed through the smooth ER to the Golgi body. Their movement along the reticulate channels is energy-dependent, since the proteins stay in the rough ER if ATP synthesis is inhibited. However, this is difficult to interpret since many inter-related chemical, osmotic and mechanical reactions depend on ATP. Palade has speculated that "lock gates" may control transport between the ER and the Golgi body. In fact, the morphological evidence for continuous channels between ER and the Golgi body is weak, and some researchers favour vacuolar shuttles between them.

The Golgi bodies are reception, finishing, packaging and despatch centres for proteins in animals, and polysaccharides and proteins in plants. Many cell secretions are in the form of glycoproteins and, although considerable glycosylation takes place in ER, the finishing touches may be a Golgi function. Extracted membranous fractions from animals, identified as Golgi elements by morphological markers, particularly intravesicular clusters of very-low-density lipoprotein markers, show high glycosyl transferase activity. These fractions are particularly rich in galactosyl-, fucosyl- and sial-transferases, enzymes that catalyze the transfer of specific carbohydrate groups to proteins. The fine structure of the Golgi body varies with cell type and physiological condition, some of the most complex being found in plants (figure 6.3). Typically it is composed of numerous stacked saccules in saucer-shaped configuration with the concave (distal or *trans*) side facing the cell surface, and the convex (proximal or *cis*) side towards the nucleus. Tubules or vesicles are found at the margins, often attached to the saccules by fine tubes (figure 6.4). These various subunits of

DIAGRAM

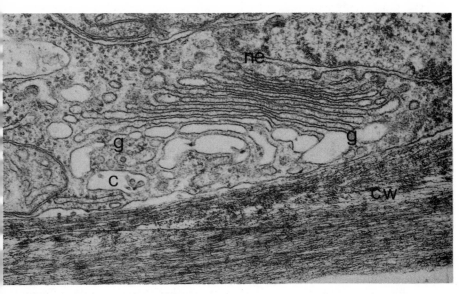

Figure 6.3 Electron micrograph of a part of a cell of a parenchymatous brown seaweed *Scytosiphon lomentaria* showing a large Golgi body (g) and its relation to the nuclear envelope (ne). The contents of the Golgi cisternàe (c) are probably material for the cell wall (cw). (Dr. A. D. Greenwood) (× 40 000).

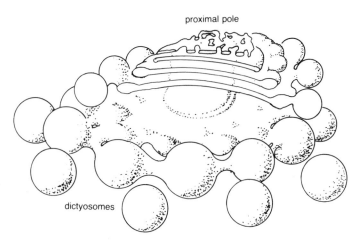

Figure 6.4 Structure of the Golgi body based on freeze-etch pictures by Fineran (1973).

the Golgi body as revealed by E.M. are sometimes called *dictyosomes*. Radioautography shows that the proteins received from the ER become concentrated in vesicles in positions that are characteristic of the cell type. In many cells there is a gradient of maturation from the proximal to the distal side of the Golgi body, giving the impression that the proximal saccules are freshly derived from ER and are displaced distally as they mature. The detailed topology of Golgi bodies, the stacking of cisternae and the degree of continuity between Golgi elements have not been worked out, and little can be said about the functional significance of the different regions. The contents of the saccules and vesicles, usually on the distal face of the organelle, become progressively concentrated by addition of new material from the sites of synthesis in the ER. There is some evidence that concentration may be due to sulphonated polyanions binding the positively charged secretory proteins into complexes. Eventually the vesicles are released from the Golgi complex as secretion vesicles or granules (figure 6.5). Immunochemical studies show that each vesicle contains a mixture of secretory proteins. This makes it unlikely that there are specific intraGolgi sites for particular secretory proteins. However Golgi bodies also produce lysosomes which contain a quite different enzyme spectrum from secretory vesicles, and this may mean some division of labour at the binding and packaging stage. In animal cells, Golgi bodies handle a variety of secretory material in addition to protein, for example, in some connective tissue cells, carbohydrates like hyaluronic acid and chondroitin sulphate are packed into secretion vesicles in the Golgi body.

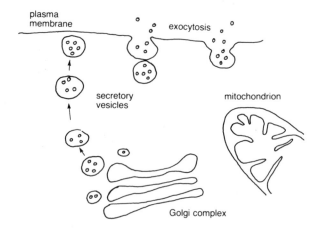

Figure 6.5   Secretory vesicle production and exocytosis.

However, the involvement of Golgi bodies in processing and transporting polysaccharides has been most clearly shown in plant cells.

6.2 The role of Golgi bodies in plant cells

In plant cells the Golgi vesicles are involved in both polysaccharide and glycoprotein biosynthesis and transport. The largest and structurally most complicated Golgi bodies have been found in algal cells such as *Pinnularia* and *Microsterias*. In higher plants, Golgi bodies are particularly prevalent in secretory cells and in young rapidly growing cells. Accumulating evidence points to the Golgi bodies in these cells being responsible for the final stages in the synthesis of macromolecules such as proteins and polysaccharides, which are transported to the plasmalemma in Golgi vesicles which fuse with the plasmalemma and discharge their contents by exocytosis. During cell division, at the end of mitotic telophase, Golgi vesicles line up in the region where the new cell plate will form, separating the two new daughter cells. When radioactively-labelled glucose is fed to these cells, it appears within the Golgi vesicles. Autoradiography of growing pollen tubes also provided clear-cut evidence that labelled precursors of pectin and polygalacturonic acid is synthesized into macromolecules in the Golgi vesicles, and transported in them to the plasmalemma and incorporated into the growing cell walls (Dashek and Rosen, 1966). Similar experiments have shown that the glycoprotein of plant cell walls which contains a high proportion of the imino acid hydroxyproline, is also assembled in Golgi vesicles and transported in them to the cell wall.

The function of the Golgi apparatus in root cells of higher plants has been carefully examined by Northcote's group and by Ray's group. The cap tissue of the roots of higher plants is extremely rapidly growing and in maize, for example, the ten thousand cells of the root cap are sloughed off and replaced every twenty-four hours. Root cap cells contain several hundred Golgi bodies with very large vesicles (figure 6.6) in which occur the final stages in the synthesis of a hydrated mucopolysaccharide containing a high proportion of fucose residues and secreted as a slime which aids the penetration of the root through the soil. ^{14}C-labelled glucose fed to the tissue appears in the slime in less than 2 minutes, and it has been calculated that the Golgi vesicles discharge their entire contents every 2 seconds. In the more mature regions of the root, where massive cell growth is taking place, the Golgi bodies are involved in the synthesis of non-cellulosic cell-wall polysaccharides which are incorporated as the wall expands. Northcote

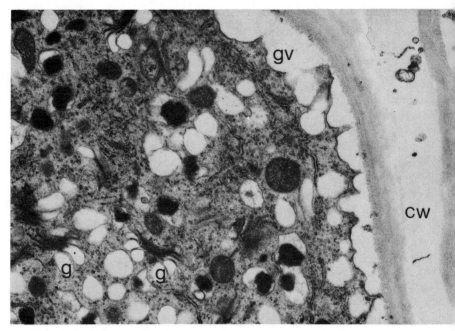

Figure 6.6 Electron micrograph of part of a root cap cell of barley (*Hordeum vulgare*) showing numerous Golgi bodies (g) with distended cisternae. The vesicles (gv) near the cell wall (cw) are involved in slime secretion. (A. Wilson) (× 15 000).

and Pickett-Heaps (1966), in a series of elegant pulse-chase experiments, fed wheat roots with 6–^{3}H-glucose for 10 minutes and sampled them for electron microscopical radioautography at intervals in the chase period. After 10 minutes chase, silver grains were present only in the region of the Golgi vesicles, and none were present over the wall. During the next 50 minutes, progressive loss of radioactivity from the Golgi bodies took place with concomitant incorporation occurring into the cell wall. Analysis of the radioactive material in the wall showed that it was a polysaccharide and that 70% of the label was now in galactosyl residues. Isolated cell fractions enriched in Golgi bodies from these tissues contain soluble polysaccharides of molecular weights greater than forty thousand, whereas fractions enriched in ER have more membrane-bound low-molecular weight polymers of molecular weight less than four thousand. There is no convincing evidence that cellulose is assembled in the Golgi vesicles; the accumulated evidence points to assembly near the plasmalemma, probably on its outer surface.

6.3 Exocytosis

Secretory vesicles produced in Golgi bodies move to the periphery of the cell, where they fuse with the plasma membrane, opening to release their contents to the exterior. This process, *exocytosis*, occurs in both plant and animal cells. It extends to secretory vesicles that are not, apparently, loaded by means of ER-Golgi systems; for example, it is the mechanism for release of adrenaline and neurotransmitters from chromaffin granules and synaptic vesicles (chapter 9). Studies of exocytosis in animal cells have been facilitated by the finding that it can be stimulated by external effectors. High glucose levels stimulate insulin secretion by pancreatic cells, acetylcholine triggers the release of epinephrine from chromaffin vesicles, and certain hormones can also induce exocytosis. A common feature of effectors is that they increase the free Ca^{2+} ion concentration at the site of exocytosis. How this is done is not well defined, but there are various possibilities, e.g. by binding to membrane phospholipids and displacing loosely bound Ca^{2+}, by acting as Ca^{2+} ionophores in the membrane, and permitting external Ca^{2+} to enter, or by inhibiting the Ca^{2+} translocating ATPase that normally maintains low levels of free Ca^{2+} by pumping out excess Ca^{2+}. Exocytosis can be induced by microinjection of Ca^{2+}, and fusion of isolated vesicles and liposomes occurs on adding Ca^{2+}. Typically membranes have a net negative charge arising mainly from the polar heads of phospholipids and the carboxyl groups of membrane proteins. Ca^{2+} neutralizes or shields these charges, thus reducing the electrostatic repulsion barrier between the membranes and allowing close contact. Fusion itself would require destabilization of opposed membranes, and it has been suggested that Ca^{2+} binding produces phase changes in the lipid bilayers that lead to transient destabilization of the membranes that makes them susceptible to fusion at their domain boundaries (Papahadjopoulos et al., 1977). A useful discussion of fusion and exocytosis is provided by Kelly et al. (1979). At present it is difficult to assess whether all instances of exocytosis are Ca^{2+} mediated, as we know relatively little about intracellular gradients and pools of Ca^{2+}, and the proportions of free and bound forms. The same problem applies to the fusion of intracellular organelles which may be regarded as special cases of exocytosis.

If new membrane is continually being added to plasma membrane during exocytosis, a similar amount must be removed at the same rate if a steady state is to be maintained. The importance of this is illustrated by the fact that the fusion of vesicular membrane into some plant cell plasma membranes during exocytosis is equivalent to adding the surface area of the cell every four minutes or less. More impressively, in crayfish motor neurones, the incorporation of vesicle membrane into the nerve terminal membrane would be enough in an axon 10 μm in diameter to account for a linear growth of 154 cm per hour! The most important mechanism for recycling membrane is probably coupled endocytosis, a process discussed in the next chapter.

6.4 Other functions of endoplasmic reticulum

Many enzyme activities unconnected with protein synthesis and modification have been found in ER fractions. These include enzyme systems that catalyze hydroxylation and desaturation reactions. The best characterized is a flavoprotein, NADPH-cytochrome P450 reductase, coupled to a membrane-bound cytochrome P450. This electron transport system catalyzes the hydroxylation of many compounds, e.g. fatty acids, steroids, amino acids and certain drugs. A postulated mechanism for the hydroxylation of progesterone to corticosterone is shown in figure 6.7.

Hydroxylation is the first step in the breakdown of drugs like barbiturates and, significantly, drug administration often induces a massive increase in smooth ER. Another ER electron transport chain is the flavoprotein cytochrome b_5 reductase and cytochrome b_5. This system is able to desaturate fatty acids, e.g. palmitic acid to palmitoleic acid (figure 6.8).

One of the most prominent ER enzymes is glucose-6-phosphatase, which catalyzes the breakdown of glucose-6-phosphate to glucose and phosphate. The functional significance of it being sited in ER is not known. Other ER enzymes include choline phosphotransferase and diacylglycerol-acyl transferase, which are involved in phospholipid and triacylglycerol synthesis. The location of these enzymes in ER correlates with the finding that smooth ER is highly developed in tissues that are particularly active in lipid synthesis; for example, adrenal cortex, a centre for steroid synthesis, has an extensive smooth ER system. This is also the case for intestinal absorptive cells which are active in synthesizing triacylglycerols from fatty acids absorbed from the intestinal lumen. The evidence generally supports smooth ER as the main centre for lipid synthesis and processing. Many of these lipids are destined for membrane synthesis and appear to be ferried to these sites in vesicles.

One of the many glycosylations that occur in the ER of liver is the conjugation of bilirubin, a breakdown product of haemoglobin, to

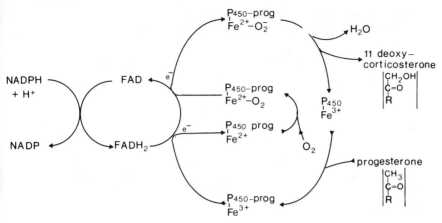

Figure 6.7 The synthesis of corticosterone occurs in endoplasmic reticulum by hydroxylation of progesterone at carbons 11 and 21. The figure shows a proposed mechanism for the hydroxylation at C11 to give the intermediate 11-deoxycorticosterone. The mechanism probably applies to all hydroxylations by NADPH-cytochrome P450 reductase and cytochrome P450 (FAD = flavin group). The progesterone is bound by the Fe^{3+} form of P450. P450 is then reduced by the flavoprotein to its Fe^{2+} form and oxygenated. After a further reduction by an electron from FADH to form an O_2^- radical, there is an internal oxido reduction to give 11-deoxycorticosterone.

bilirubin glucuronide. The glycosylation makes it possible for bilirubin to be secreted into the bile and subsequently degraded in the intestine.

6.5 Biosynthesis and assembly of endoplasmic reticulum

The foregoing has shown that ER is a complex highly adaptive structure that increases in size in response to drugs and toxins. Neither the structure

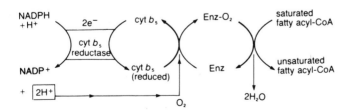

Figure 6.8 Desaturation of palmitoyl CoA by the NADPH-cytochrome b_s system of the endoplasmic reticulum. In this process 1 molecule of oxygen is reduced by 2 pairs of electrons, one pair from NADPH and the other from the unsaturated fatty acyl CoA.

of ER nor its mode of assembly and disassembly is well defined. However, the rate-limiting factor for its growth appears to be the rate of production of protein, and presumably the control mechanisms are at the gene transcription or mRNA translation level. Protein is inserted directly into the ER membranes from bound ribosomes, and the main glycosyl transfer enzymes operate at the cytosol side or even on the ribosomes. This is in contrast to glycosylation of secretory proteins which occurs on the lumen side. The significance of a high proportion of the structural and catalytic proteins of the ER being glycosylated is not clear. Probably the sugar residues are fundamental to the orientation and location of the proteins in the membrane. There are no obvious growth points, and proteins are inserted over large areas of ER and show considerable lateral movement within the membrane. The subject is reviewed by Dallner et al. (1979).

CHAPTER SEVEN

LYSOSOMES

7.1 Lysosomes

Cell biology is a continuing saga of controversy, speculation and progress. Having adopted a strategy of divide and conquer, the researchers of the 1940s and 50s were faced with the task of defining the cell fractions the new techniques had produced. Amidst controversies that seem strange in retrospect (e.g. whether mitochondria originated from microsomes or vice versa!) a new organelle was discovered—the lysosome (figure 7.1). The discovery was made in 1955 by de Duve who observed that the activity of acid hydrolases in animal tissue homogenates increased with time, and proposed that these enzymes were present inside vesicles that became leaky with age. In face of arguments that the enzymes had probably leaked out of damaged mitochondria, de Duve maintained that a new organelle was involved, which he named the lysosome to indicate that the internal enzymes only became apparent when the membrane was lysed. For this and a brilliant series of experiments on lysosomes, de Duve shared the 1974 Nobel Prize for Physiology with Palade, of endoplasmic reticulum fame, and Claude, another cell biology pioneer.

The existence of the lysosome was confirmed by two experimental approaches. First, the main intracellular sites of acid hydrolase activity were determined cyto-chemically and, by application of fluorescent antibodies against these enzymes to tissue slices, and shown to be in membrane-bound vesicles. The vesicles were heterogeneous and pleomorphic, and not markedly different from other cytoplasmic vacuoles, except for the presence of an electron-lucid rim or halo just inside the membrane, that helps identification by E.M. The second approach was to improve the methods of isolating lysosomes, since the original de Duve fractions were contaminated with mitochondria, microsomes and microbodies. In the 1960s it was found that rats injected with dextran or Triton WR incorporated these into their lysosomes, thereby altering their density and making cleaner separation possible by differential centrifugation and density gradients possible. However, these lysosomes

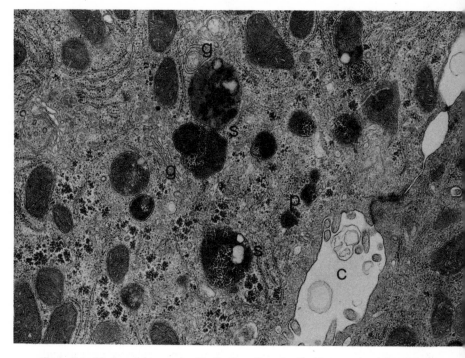

Figure 7.1   Electron micrograph of isolated rat liver that had been perfused for 1½ hours, showing primary lysosomes (p) and the development of secondary lysosomes (s) after fusion with endocytotic vesicles. The section shows Golgi bodies (g) ord bile canalicula, the main source of the endocytosed material (Dr. Gillian Bullock) (× 15000).

are atypical in some respects, and a more favoured current method is separation of the crude lysosome fraction by continuous electrophoresis. Analysis of lysosomal fractions, some of doubtful status, has identified over sixty putative lysosomal enzymes. Some of the best authenticated are shown in table 7.1, and it is obvious that as a group they are capable of degrading almost any biological polymer—proteins, carbohydrates, glycoproteins, glycolipids, lipids, nucleic acids. They include enzymes that can catalyze the hydrolysis of glycerol esters, various classes of phosphate and sulphur esters, glycosyl bonds and peptide bonds. There is evidence for qualitative and quantitative differences in the enzyme content of lysosomes in different tissues, and sometimes even in the same cell.

Table 7.1 Major enzyme activities associated with lysosomes

Enzyme	Active against
acid lipase phospholipase A1 and A2 phospholipase C	LIPIDS
acid DNAse acid RNAse	NUCLEIC ACIDS
acid phosphatase acid pyrophosphatase phosphodiesterase	
N-acetyl galactosamidase N-acetyl hexosaminidase glucuronidase hyaluronidase neuraminidase chondrosulphatase sulphaminidase	COMPLEX LIPIDS, POLYSACCHARIDES, and GLYCOPROTEINS
acid carboxypeptidase cathepsins A, B, C and D. amino acid naphthylamidase	PROTEINS

7.2 Lysosome synthesis

Lysosomal enzymes appear to be synthesized on rough ER and discharged into the ER cisternae, through which they pass to the Golgi complex, where they are packaged into primary lysosomes. The evidence for this is based mainly on the separation and analysis of microsomes, Golgi vesicles and lysosomes at different times after stimulation of lysosome production by gonadotrophin, and on experiments that follow the cellular distribution of pulse-labelled lysosomal glycoproteins, by autoradiography. It is thought that in some cells, namely neurones and exocrine pancreas, lysosomes arise from enzyme-rich ER regions without direct Golgi involvement. In fact, there may be various routes and mechanisms for transport and packaging of lysosomal enzymes with one or other being dominant in different cells (Holtzman, 1975). Some idea of the biochemical basis of how ER and Golgi segregate lysosomal enzymes into lysosomes and proteins for export into secretory vesicles, is now emerging. It appears that when lysosomal enzymes are synthesized into the ER they become glycosylated to much the same extent as secretory proteins. However at some, as yet undefined, region, presumably near the packaging site, the glycosyl units of the lysosomal enzymes are "clipped down" by glycosyl hydrolases, leaving a phosphorylated mannose that acts as a tag for binding the enzyme to the

140   LYSOSOMES

membrane. (All lysosomal enzymes seem to be membrane-bound, although the strength of binding varies from one enzyme to another.) The simplest model of subsequent events is that, when a particular density of enzyme binding is achieved, the membrane in that region becomes destabilized and forms a primary lysosome or a vesicle that will finish up as a primary lysosome.

There is some evidence for spatial separation of lysosome formation and the formation of other types of vesicles in the Golgi of rabbit leucocytes. However, the whole issue of packaging sites and how molecular species are

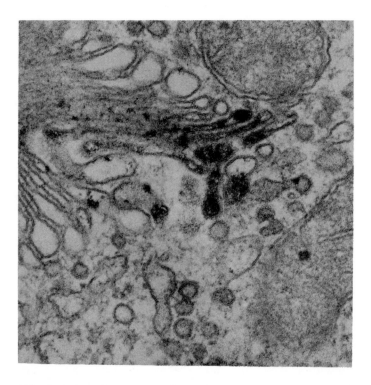

Figure 7.2 A section of neuronal perikarya from larval frog spinal cord incubated histochemically to show thiolacetic esterase (lysosomal marker enzyme) activity—the darkly staining areas. This provides evidence that lysosomes are formed in the region of the Golgi apparatus. Serial sectioning suggests that in this and certain other tissues lysosomes are not simply packed and budded off from the Golgi, but that the enzymes may pass directly from rough ER into Golgi associated smooth ER from which lysosomes are formed. Provided by Dr. R. S. Decker (× 99 000).

directed to them is difficult, and few firm statements can be made at present. Research into lysosome packaging has been assisted by studies of inherited lysosomal storage diseases, particularly "I cell" disease. This disease is characterized by intracellular accumulations of various macromolecules, because the lysosomal hydrolases normally responsible for breaking them down are leaked into the lymph rather than retained in the cells. It now seems that the disease is due to a mutation of the glycosyl hydrolase enzymes that prevents them from "clipping down" the glycosylated pro-lysosomal enzymes to the forms that bind to the Golgi membrane. Consequently, instead of becoming bound inside lysosomes, the enzymes remain in proenzyme form and are secreted like secretory proteins. In addition to providing a means of intraGolgi segregation, the binding properties of lysosomal enzymes are important in two other respects: bound enzymes are not lost when lysosomes fuse, as they sometimes do, with plasma membrane, but are retained in the plasma membrane and available for recycling. Also, intralysosome binding probably holds the enzymes in a conformation that prevents the active centres attacking the membrane and one another.

7.3 Lysosome function

Primary lysosomes carry out several functions, the relative importance of which varies from tissue to tissue. They break down foreign material, degrade transport proteins, so that their contents become available to the cell, digest effete organelles and excess macromolecules produced by the cell, degrade extra-cellular material, and sometimes play a part in programmed embryonic development by destroying tissue.

Degradation of foreign material
External material that cannot permeate the plasma membrane by other means often enters cells by endocytosis, a process whereby the material is engulfed by plasma membrane and brought into the cell enclosed in vesicles (figure 7.3). The endocytosis of particulate material is termed phagocytosis and is a feature of many cell types; for example, protozoans like *Amoeba* and certain cells of digestive tracts in lower animals depend on phagocytosis for the major part of their food supply; polymorphonuclear leucocytes in blood phagocytose invading microorganisms, and macrophages of the reticuloendothelial systems of liver, spleen and other sites clear the blood of effete erythrocytes and tissue fragments by phagocytosis. The physicochemical basis of phagocytosis is not

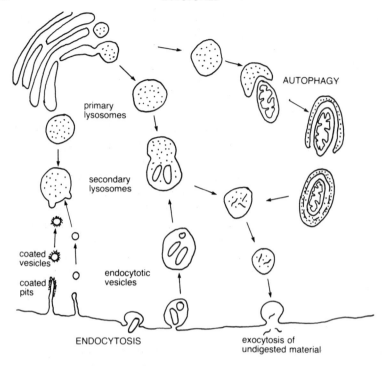

Figure 7.3   Diagram of lysosomal involvement in intracellular degradative processes.

understood. The killing or denaturation of bacteria and viruses begins at the cell surface where they are exposed to the toxic agents hydrogen peroxide (H_2O_2) and superoxides (O_2^-), apparently produced by the action of a membrane-bound NADPH dehydrogenase. This bacteriocidal activity goes on even after the microorganisms have been enclosed in phagosomes, the cytoplasm of the cell being protected by superoxide dismutase, an enzyme that breaks down superoxides. Phagosomes fuse with primary lysosomes to form hybrid vesicles called *secondary lysosomes* within which digestion takes place. The low-molecular-weight products diffuse out through the lysosome membrane into the cytosol and are recycled. Indigestible material in secondary lysosomes can be released from at least some cell types by exocytosis; this is particularly the case for protozoans.

   The endocytotic uptake of soluble material is termed *pinocytosis* and results in much smaller vesicles that can be clearly defined only by E.M.

Pinocytosis probably occurs in all animal cells and is the mechanism for uptake of proteins, antigen-antibody complexes and soluble macromolecules of various kinds. Although much of the material that is pinocytosed is taken up non-selectively, the uptake of many substances is highly specific, carefully governed and of great physiological importance. The best researched example of this is the uptake of LDL, low-density lipoprotein that carries cholesterol in the plasma from the liver to the rest of the body (reviewed by Goldstein *et al.* 1979). Each LDL particle consists of an internal core of 1500 or so cholesterol molecules linked to long-chain fatty acids surrounded by a phospholipid/protein coat. The cell surface has up to 15 000 binding sites for LDL, detectable by following the binding of LDL-ferritin or radioactive LDL. These sites or receptors are clustered in fine indentations of the membrane called *coated pits,* because short fibres radiate from the pits into the cytosol. The number of receptors varies with the cell's requirement for cholesterol, a control device that prevents cholesterol overloading (Goldstein & Brown, 1977). LDL binding is followed within 5–10 minutes by the pinching off of pit membranes into coated vesicles containing the LDL. These vesicles eventually fuse with primary lysosomes to form secondary lysosomes (figure 7.3) within which the LDL is degraded and the cholesterol molecules cleaved from their fatty acids and made available to the cell. The LDL receptors are conserved and recycled. Other carrier particles enter cells by similar systems, e.g. transcobalamin II undergoes receptor mediated endocytosis and is degraded in the lysosomes to release vitamin B 12.

Most endocytosed proteins finish up in lysosomes, but there are significant exceptions such as maternal immunoglobins which are transferred across neonatal intestinal epithelial cells in coated vesicles and discharged into the neonatal circulation. Also, yolk proteins taken up by oocytes accumulate in yolk granules without lysosomal degradation. Evidently some classes of vesicles are less prone than others to collision or fusion with lysosomes. Some form of cytoskeletal segregation, e.g. microtubule guide tracks cannot be discounted, but it seems likely that the main factor that governs the interaction of vesicles with specific membranes is the nature of the vesicle membrane. This could be a function of the ligand and the type of receptors in the pits from which the vesicles were derived. Coated vesicles are made mainly of lipid and a protein called clathrin (MW 180 000) which is also present in pits, but little is known of the molecular organization of pits or the transformations that occur when vesicles are formed.

Some low-molecular-weight materials like drugs and dyes enter lysosomes by diffusion through the membrane and are often complexed inside and degraded with difficulty or not at all. It has been suggested that accumulated material that damages lysosomes may cause malignant transformation of cells by bringing about leakage of lysosome enzymes that attack the genetic material in the nucleus. A potentially useful method of treating cancer, making use of lysosomes, involves enclosing anti-mitotic drugs in tiny liposomes (artificial vesicles) and perfusing them through regions of tumour formation. There is some evidence that the liposomes are taken into the lysosomes of tumour cells faster than those of normal cells, because the former have faster rates of pinocytosis. After digestion of the liposome carrier, the drug diffuses out of the lysosome and stops cell division.

Degradation of internally produced material
Lysosomes are able to digest intracellular structures including mitochondria, ribosomes, peroxisomes and glycogen granules. This is particularly evident in cells undergoing programmed death, but it also occurs to some extent in apparently normal cells. This process, termed *autophagy,* takes several forms. In some cases the lysosome appears to flow around the cell structure and fuse, enclosing it in a double membrane sac, the lysosome enzymes being initially confined between the membranes (figure 7.3). The inner membrane then breaks down and the enzymes are able to penetrate to the enclosed organelle. In other cases the organelle to be digested is first encased by smooth ER, forming a vesicle that fuses with a primary lysosome.

Autophagy raises some interesting questions. What induces lysosomes to phagocytose their own cell organelles in programmed cell death? We don't know. Cell death is accompanied by changes in the cell oxido-reduction and osmotic potentials, and often acidification takes place as a result of anaerobic glycolysis of glycogen. Changes like these could conceivably change the charge patterns of the surfaces of cell structures and allow them to interact with lysosome membranes. Another factor is that the Ca^{2+} concentration in the extracellular fluid is normally higher than in the cells, and irreversible damage to the membrane would tend to increase the intracellular Ca^{2+}. This might also influence autophagy, taking account of the apparent link between Ca^{2+} and membrane fusion phenomena discussed in the last chapter. At least the necessity for autophagy in programmed cell death in developmental processes is understandable; there must be rapid and effective excision of tissue. Autophagy in seemingly normal cells is more difficult to comprehend. Is it a means of clearing out the occasional effete or malconstructed organelle? If so, how are these recognized? Is it a process to correct overproduction of organelles? Is it a response to environmental conditions that are best met by redistribution of cell resources? Are occasional rouge lysosomes produced? This is an interesting area of cell biology with more questions than answers.

Under some conditions, cells degrade their constitutive proteins. Starvation results in the breakdown of muscle protein for fuel; cells sometimes replace one group of proteins by another in response to changing environment, and there is evidence that abnormal proteins can be recognized and degraded. Unfortunately, much of the work in this area is difficult to assess because of uncertainties over the purity of the various membrane and organelle fractions used in *in vitro* work. However, lysosomes appear to be one of the systems involved. Some proteins show a very short life and a high turnover rate, and it is unlikely that autophagy is involved in their breakdown, as the process would be too slow. However, the kinetics of the degradation of longer-lived proteins in various cell types is compatible with autophagy, and their breakdown is reduced by inhibitors that prevent vacuole formation. This suggests that soluble internal proteins can be packaged into vesicles or vacuoles that fuse with the lysosomes. Although this may explain how large proteins are brought into contact with lysosome enzymes, it raises the problem of how such proteins could be packaged, particularly on a selective basis. The importance of lysosomes in the degradation of other materials produced in the cell is illustrated by the finding that genetic deficiencies in specified lysosome enzymes can give rise to intracellular accumulation of glycogen, mucopolysaccharides, glycolipids and lipids. This underlines the role of lysosomes in maintaining a steady state in cells.

Extracellular digestion
Extracellular degradation is sometimes induced by lysosome enzymes exocytosed at the cell surface. The best authenticated case is the exocytosis of Cathepsin D by fibroblasts, and its role in the breakdown of connective tissue preceding ossification (Dingle, 1973). Acid hydrolases are released from osteoclasts and are important in breaking down bone for reabsorption; these cells also secrete lactic acid which makes the local pH acid enough for optimal enzyme activity. The enzyme elastase is secreted concomitantly with the rapid phagocytosis that occurs in the buccal cavity during mastication. Lysosomal hydrolases are released into seminal fluid. These and other instances where lysosomal enzymes are secreted for a physiological function raise many questions. What triggers their release? Are they bound less firmly to lysosomal membranes than other enzymes, so

that they readily dissociate when lysosomes fuse with the plasma membrane? Are they chemically modified by, for example, glycosylation at the cell surface, so that they are not endocytosed back into the cell?

7.4 The internal environment of the lysosome

Lysosome membranes are characterized by substantial amounts of carbohydrate material, particularly sialic acid. Most of the hydrolase enzymes are membrane-bound, which may prevent the active centres of the enzymes gaining access to susceptible groups in the membrane. Most lysosomal enzymes have acid pH optima and, significantly, the intralysosomal pH seems to be about 2 pH units lower than that of the cytosol, as judged by the distribution of labelled lipid-soluble weak acids across the membrane and titrations of intact and broken lysosomes. Shortly after fusion between primary lysosomes and phagosomes, the pH of the secondary lysosomes drops to the mean optimum for the lysosomal enzymes. It has been proposed that this is due to a membrane-bound H^+ translocating ATPase driving H^+ into the lysosome. However, although bicarbonate-stimulated ATPase activity has been demonstrated in lysosome preparations, the evidence for a proton pump is not compelling. An alternative explanation is that the low intralysosomal pH is caused by a quasi-Donnan equilibrium generated by the presence of glycoproteins with low isoelectric points. These would act as fixed anions and create acid conditions. The state of information on the pH and transport of protons in lysosomes was reviewed by Reijngoud (1978). A practical consequence of the low pH optimum for lysosomal enzymes is that the effects of enzymes leaked into the cytosol through lysosome disruption are likely to be less serious than if they showed optimal activity at the same pH as the cytosol.

7.5 Lysosome-like bodies in plant cells

No particles which strictly conform in all respects to animal lysosomes and exhibit the combination of morphological and biochemical properties diagnostic of them have yet been identified in plant cells. But in plant cells a heterogeneous group of vesicle-like particles has been recognized, distinct from mitochondria and Golgi vesicles, and exhibiting some of the hydrolytic and autophagic properties characteristic of animal lysosomes. Almost certainly these particles are pleomorphic and show changing morphological and biochemical characteristics, as the cells in which they are present develop or senesce. These dynamic changes and the differing functions of the particles with similar structures in different tissues account for the difficulty in assigning them to a tightly-defined category of particle.

Many plant cells contain approximately spherical particles approximately 1 μm in diameter, bounded by a single membrane which can be stained cytochemically for hydrolytic enzymes such as acid hydrolases and acid phosphatases. In most studies, no latency properties as are characteristic of animal lysosomes have been assigned to these particles. Ultrastructurally, these particles show a variety of morphological characteristics: some resemble small vacuoles and appear to engulf mitochondria and cell membranes; others present in germinating seeds such as castor bean resemble spherosomes and initially contain 90% of the lipid of the endosperm (as tri-glyceride) which is mobilized during germination. There have been several attempts to isolate these lysosome-like particles from plant cells and characterize them biochemically, notably by Matile's group. Vacuoles, 0·3–1·5 μm in diameter, can be co-sedimented with mitochondria from homogenates of seedlings, and such preparations exhibit high activity for protease, phosphatase, esterase and ribonuclease. Aleurone grains from seeds such as peas have also been shown to contain a wide range of hydrolytic enzymes, including proteases and phosphatases involved in the mobilization of the protein reserve in the endosperm cells. It seems possible, although not proven, that the small vacuoles in developing plant cells may well be responsible for the autophagic function assigned in animal cells to lysosomes.

CHAPTER EIGHT

ORGANELLES IN SPACE AND TIME

8.1 Integration of metabolic processes

The purpose of this chapter is to outline some examples of how the division of labour associated with cell compartmentation is integrated, and to consider ideas on how the eukaryotic cell may have arisen. The interdependence of processes in different cell compartments has been recognized for many years: inhibition of DNA synthesis and transcription in the nucleus quickly stops protein synthesis in the cytoplasm, leading to breakdown of cell structure and death. If photosynthesis in green cells is stopped, neither the units for biosynthesis nor the substrates for respiration in the mitochondria are produced and the cells degenerate. Addition of uncouplers or inhibitors of oxidative phosphorylation to animal cells results in insufficient ATP for active transport, biosynthesis and electrical or motor activity, with predictable results. Insult to lysosomes releases acid hydrolases that degrade other cell constituents. From recognizing the basic facts of interdependence of compartments we have moved in recent years to some understanding of their integration and control.

It was mentioned in chapter 1 that the mitochondrial membrane was impermeable to NAD^+ and NADH, and that this made it possible to have different $NADH/NAD^+$ ratios in the cytosol and the matrix. But how is the NADH that is continually generated in the cytosol oxidized if it does not have access to the respiratory chain? Figure 8.1 shows one way. The NADH is oxidized by oxaloacetate in the cytosol, and the resulting malate carries the hydrogen atoms into the mitochondrion, where they eventually enter the respiratory chain. A complex shuttle system is used involving anion carriers and cytosol and mitochondrial enzymes. The significance of different redox pools is evident from a consideration of the optimal conditions for glycolysis in the cytosol and oxidative phosphorylation in the mitochondria. The reaction below is a key step in the production of pyruvate:

$$3\text{-phosphoglyceraldehyde} + NAD^+ + Pi \rightleftharpoons 3\text{-phosphoglyceroyl phosphate} + NADH + H^+$$
$$\Delta G^{0\prime} = +7 \text{ kJ mol}^{-1}$$

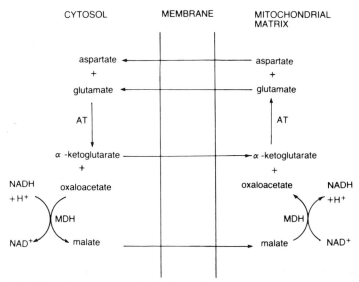

Figure 8.1 The malate-aspartate shuttle for making reducing equivalents from cytosolic NADH available to the respiratory chain. MDH = malate dehydrogenase. The shuttle is depicted as a one-way system for clarity. It is completely reversible. The anion translocases involved have not been specified because, as figure 4.21 shows, there is considerable flexibility with respect to counter anions. AT = aminotransferase.

Typically the concentration of 3-phosphoglyceraldehyde in the cytosol is very low, because of the thermodynamic poise of preceding reactions in the sequence. Consequently a high $NAD^+/NADH$ ratio in the cytosol (about 10^3) is necessary to drive the reaction to the right in the direction of pyruvate for oxidation in the TCA cycle. However, in the mitochondrion the main oxidizing agent for pyruvate and other substrates is NAD^+, the resulting NADH being oxidized via the respiratory chain (figure 4.11) This is not possible if the mitochondrial $NAD^+/NADH$ ratio is as high as the 10^3 optimum for glycolysis. In fact if the $NAD^+/NADH$ ratio was much greater than 10, the electrons would tend to move up the chain and reduce NAD^+ in an energy-dissipating process rather than down the chain from NADH in an energy-yielding process. Happily separation of glycolysis and oxidative phosphorylation into two physically separated compartments with different $NAD^+/NADH$ has solved this problem.

Continuing the theme of energy metabolism, the respiration rate of many tissues increases when the cells work faster, and decreases when the rate of working is experimentally reduced, e.g. when the ATPase that pumps Na^+ and K^+ across the membrane of kidney cells is inhibited, the respiration rate decreases by about 50%. How does the operation of a work site on the plasma membrane influence the rate of electron transport in mitochondria? The main control factor is the (ATP)/(ADP)(Pi) ratio. When this ratio is high, as it tends to be in quiescent cells not using ATP in large amounts, mitochondrial respiration is kept in a controlled state (figure 4.15). However, when cellular work occurs, ATP is used faster and the ratio decreases to a lower steady state, thus releasing electron transport from its controlled state. Changes in respiration of cells are never as marked as those noted for isolated mitochondria in figure 4.15, where the experimental conditions are precisely controlled. The adenine nucleotide ratio also influences the rate at which pyruvate becomes available to the mitochondria. High levels of ATP, and low ADP and AMP in the cytosol, cause allosteric inhibition of phosphofructokinase (PFK), thus reducing the rate of pyruvate formation. If cell work significantly reduces the cytosol ATP levels and increases ADP and AMP, the inhibition is released and pyruvate is produced faster. This is one of several control mechanisms that ensure that glucose and glycogen are not degraded to oxidizable intermediates when the cellular demand for oxidative phosphorylation is low. There is also the intriguing question of the control of the balance between photosynthesis and respiration in green cells in the light of which the translocases of the chloroplast envelope are capable of facilitating the movement of intermediates of the two processes in and out of the chloroplast at rates ten-fold higher than the maximum rate of photosynthesis. The phosphate translocator allows movement of dihydroxyacetone phosphate (DHAP) and 3-phosphoglycerate (PGA) at rates in excess of 1800 μatoms C/mg chl/hr. The flooding of the cytosol with respiratory substrates would be expected to lead to increased respiration in the light, since resultant changes in the [ATP]/[ADP] [Pi] ratio will increase both PFK activity and mitochondrial oxidation. Experimentally, the reverse is found to be the case: citrate is far more slowly labelled from $^{14}CO_2$ in the light than in the dark, and respiration in photosynthesizing cells appears to be almost totally inhibited. The control is exerted by the maintenance of stable differences in the redox state of pyridine nucleotides at both sides of the chloroplast envelope, both by the operation of light-induced ΔpH across the inner envelope membrane and by the oxaloacetate /malate shuttle. Both the decrease in hydrogen ion concentration in the stroma on illumination and the movement of malate into the chloroplast can lead to an increase in reduction of NAD^+, as illustrated by consideration of the equilibrium condition of the malate/oxaloacetate (OAA) shuttle (net flux = 0):

$$\left[\frac{(OAA)(NADH)(H^+)}{(malate)(NAD^+)}\right]_{stroma} = \left[\frac{(OAA)(NADH)(H^+)}{(malate)(NAD^+)}\right]_{cytosol}$$

Similarly the OAA/malate and the DHAP/PGA shuttles operating together in the light can maintain a low ATP/ADP ratio inside the chloroplast side by side with high ATP/ADP ratio outside. In the dark, the shuttle systems reverse direction and supply the chloroplast with ATP and remove excess reducing equivalents (figure 8.2) as before.

Another example of sophisticated chloroplast/cytosol interaction in the green cell is the control of starch synthesis by the concentration of inorganic phosphate [Pi] in the cytosol. Triosephosphate molecules in the chloroplast can be utilized in a variety of ways: some are required in the regeneration of ribulose bisphosphate (RubisP), others are used in starch synthesis, but in normal photosynthesizing conditions most of the triosphosphate is exported to the cytosol in counter-exchange for Pi released during the cytosolic formation of sucrose. Conditions which lower the concentrations of Pi in the cytosol result in increased starch synthesis, either naturally (e.g. by lower sucrose synthesis or in phosphate-deficient plants) or artificially (e.g. by sequestration of Pi as mannose-6-phosphate following feeding of the leaves with mannose). Clearly reduction in Pi in the cytosol results in lack of exchange of DHAP in the cytosol, and its

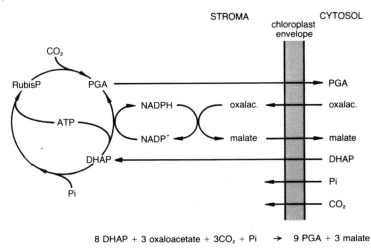

8 DHAP + 3 oxaloacetate + 3CO₂ + Pi → 9 PGA + 3 malate

Figure **8.2**   Diagram illustrating the reduction of CO_2 in chloroplasts in the dark facilitated by the oxaloacetate/malate shuttle.

conversion to starch in the chloroplast. The different control situations are illustrated in figure 8.3.

8.2 Plasticity and origin of organelles

With the possible exception of the nucleus, the organelles in many cell types show considerable plasticity in terms of shape, size and quantity. The mitochondrial content of lower eukaryotes, in particular, changes in

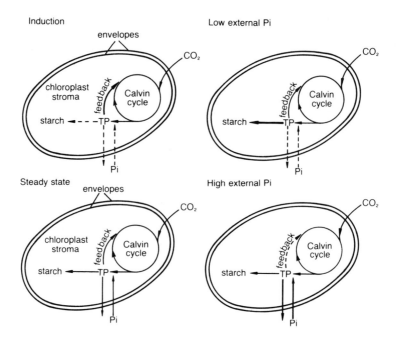

Figure 8.3 Proposed relationship between induction, Pi and metabolite movement in the chloroplast. In normal concentrations of external Pi, TP (triose phosphate) formed in the photosynthetic carbon cycle will either (a) feed back into the cycle, (b) undergo further conversion to starch, etc., on the stroma, or (c) be exported. At the onset of illumination, the lag which is observed will reflect the time taken for autocatalytic feedback to raise the concentration of cycle intermediates to the steady-state level dictated by the prevailing light intensity, etc. During this period, feedback will be favoured by rapid utilization within the cycle. As the steady state is approached, relative surplus will increasingly permit export and internal storage. In low external Pi (which it is believed can be achieved experimentally in some leaf tissue by feeding mannose) export will be diminished, and surplus TP diverted into starch synthesis, etc. In high external Pi, obligatory export will diminish the proportion of TP available for feedback and prolong induction beyond the normal period of 1–3 min (from Walker (1976) in *The Intact Chloroplast*, ed. J. Barber. Elsevier, pp. 235–278).

response to nutritional and physical conditions. Chloroplasts are also very plastic organelles; their shape is constantly changing and they can lose vesicles containing stroma from their outer surfaces. The stacking of the thylakoids also readily responds to environmental stress; nutritional depletion leads to unstacking and vesiculation of the thylakoids, while low light intensities, as is common for shaded leaves, give rise to large granal stacks with many thylakoids. Starch and plactoglobuli in plastids are temporary stores of carbohydrate and lipid which can be rapidly mobilized. The ER of many cells undergoes dramatic changes during development both with respect to overall content and relative proportions of rough and smooth ER. Administration of barbiturates and other drugs, and even mechanical injury to tissues, causes proliferation of ER. The Golgi bodies vary with the metabolic state of the cell and may be present during only one phase of the life cycle of an organism; another example is the population of glyoxysomes in the endosperm cells of germinating fatty seeds (chapter 6) which are highly active for only 7–10 days. Lysosomes wax and wane, and the populations of cell vesicles vary with nutritional conditions. The organelle complement of cells should therefore be regarded as a dynamic rather than a static organization that is able to respond to changing conditions in an adaptive manner. Although physico-chemical mechanisms are probably the immediate cause for some types of organelle flux, e.g. the fusion of primary lysosomes with endocytotic vesicles to form secondary lysosomes, exocytosis of secretory vesicles and coalescence of mitochondria, the organelle complement is determined basically by genetic activity. Organelles are dependent on the transcription and translation of the genes for the structural and catalytic proteins of which they are composed, and the enzymes concerned in their biosynthesis. Typically, the induction or repression of gene products occurs at the transcription level, and it is reasonable to suppose that most factors, defined or undefined, that influence the biosynthesis of organelles act at this point. In fact, the core of the problem of how organelles are held in dynamic equilibrium, how they respond to changing conditions, and how they differentiate during embryogenesis is genetical.

The origin of organelles has been the subject of much speculation and debate. Clearly there are many biologists who agree with Aristotle's view that a definition of an object is incomplete if you cannot tell where it came from. Taking the nucleus first, we simply do not know. The advantages of a nuclear membrane in eukaryotes have been discussed in chapters 1 and 2, but its phylogenetic origins are obscure. In the case of prokaryotes, the genome is attached to the plasma membrane, and the daughter genomes are separated by localized membrane growth. Consequently, it has been suggested that the nuclear membrane may have developed as an ingrowth of the plasma membrane. Many lower eukaryotes, particularly dinoflagellates, show membrane-mediated processes for genome separation that are prokaryote-like, and a tendency can be traced from this to mitotic-spindle-dominated segregation in the higher eukaryotes. Interested readers should consult Kubai (1975) and Fuge (1977).

The main thesis on the origin of chloroplasts and mitochondria is that they evolved from primitive types of blue-green algae and aerobic prokaryotes respectively. These are considered to have entered into symbiotic associations with primitive heterotrophic cells that hitherto had obtained their energy from the glycolytic breakdown of polymers—an ancient but inefficient process. The evidence is circumstantial and based mainly on some striking similarities between chloroplasts and extant blue-green algae, and between mitochondria and extant aerobic bacteria. Both types of organelles resemble prokaryotes in possessing DNA and a complete transcription and translation system, including ribosomes of the same-size order and sensitivity to the same drugs. The thylakoid membranes of chloroplasts are organized on the same basis as that of the single thylakoid of the blue-green alga—they both contain photosynthetic pigments (including chlorophyll a) and vectorially-arranged electron carriers and ATP synthetase enzymes. In both cases, photo-phosphorylation apparently occurs by a chemiosmotic mechanism. Similar parallels exist between mitochondria and certain aerobic bacteria. *Micrococcus denitrificans* has a very similar electron transport chain to mitochondria that is orientated in its plasma membrane, so that electron transport drives H^+ out and establishes transmembrane $\Delta\mu_{H^+}$ (up to 200 mV). In extent aerobic bacteria, such transmembrane potential differences drive ATP synthesis and substrate and ion uptake as in mitochondria. Therefore, taken together with the fact that the organelles are in the same size order as prokaryotes, there is a plausible case for their endosymbiotic origin. This is strengthened by the discovery of several contemporary endosymbiotic relations between blue-green algae and protozoans.

Assuming that the endosymbiot hypothesis is broadly correct, substantial changes must have taken place in both the original host cell and the endosymbiots. The latter have almost certainly lost most of the genes they possessed as free-living organisms, and the former have acquired new genes to provide the various species of enzymes and other proteins presently found in organelles but not coded by organelle DNA. Gene transfer between endosymbiots and hosts may have played a part in this.

This view is consistent with the smaller number of mt tRNAs coded by mt DNA, as distinct from nuclear genes in higher cells compared with yeast and the fact that, although the ATP synthetase proteolipid is coded by mt DNA in *S. cerevisiae*, it is nuclear-coded in *Neurospora crassa*. Although the involvement of two genetic systems in organelle biosynthesis and function looks more complicated than if this were monocontrolled, it is consistent with the endosymbiot theory. Once nutritional dependency is established between an endosymbiot and its host, selection might be expected in favour of processes that link them further, so that their individual growth rates and metabolisms are more closely integrated. The extent of this integration in the case of chloroplasts and mitochondria is illustrated by the examples of metabolic interdependence mentioned earlier. The metabolic consequences of the joint expression of organelle and nuclear genomes in the synthesis of two different subunits of the same protein are also illustrated by the recent demonstration in tobacco species (Kung and Rhodes, 1978) that the addition of a single polypeptide to one of the eight small subunits of ribulose biphosphate carboxylase reduces both the carboxylase and oxygenase activities of the enzyme. This shows dramatically how a change in coding information of the nuclear genome can control the metabolic function of the chloroplast. The enzyme (RubisP carboxylase) is composed of eight large subunits coded by the chloroplast genome and containing the catalytic site, and eight small subunits coded in the nuclear genome and containing the regulatory site. The metabolic consequences of change in the joint expression of chloroplast and nuclear genomes are therefore potentially very considerable.

Now that techniques of molecular genetic analysis are available to study these interactions, we can expect many more to be described. The dual control of RubisP carboxylase in chloroplasts, and cytochrome oxidase and ATP synthetase in mitochondria, effectively means that whatever these organelles may have been in the past, they no longer have any degree of autonomy of bioenergetic function. The degree of integration of the organelles with the rest of the cell is so sophisticated that, if the endosymbiot theory is correct, numerous fundamental changes must have occurred involving not just loss but acquisition of functions (e.g. adenine nucleotide translocases and acyl-CoA synthetases) not found in extant prokaryotes. Despite the problems inherent in the hypothesis, it is more logically consistent than other suggestions. Interested readers should read Broda's (1975) review. Unfortunately, unlike questions about the activities of present-day organelles, questions concerning the origin of prehistoric organelles are not amenable to direct experimentation.

CHAPTER NINE

ORGANELLES AND NEUROMUSCULAR ACTIVITY

IN THE FOREGOING CHAPTERS ORGANELLES HAVE BEEN DISCUSSED MAINLY IN terms of their intracellular roles. However, some organelles have functions that reach beyond the boundaries of the cell. In particular, it seems that in the evolution of multicellular organisms, the ER and Golgi systems have extended their spheres of influence and acquired specialized functions that are vital to the coordination of the organism. This is shown in the production and secretion of insulin via the ER-Golgi-vesicle system of the cells of the pancreas; similarly thyroxine, oxytocin and other polypeptide hormones that regulate the physiology of organs and organisms are secreted from their sites of synthesis by vesicles that appear to be a simple extension of the primitive function of ER and Golgi bodies, i.e. to deliver membrane material to the cell surface. Rapid internal communication and responsive reflex action are brought about by the neuromuscular system, the activity of which is also dependent on specialized cell organelles that are associated with ER and Golgi bodies, namely the synaptic vesicles of neurones and the sarcoplasmic reticulum of muscle. These organelles, together with the chromaffin granules of adrenal medulla, are intrinsically interesting, central to an understanding of neurophysiology, and demonstrate specialization of organelles in the service of communication between tissues.

9.1 Chromaffin granules

The adrenal medulla is a major site of synthesis of the catecholamines epinephrine (adrenaline) and norepinephrine (noradrenaline). These compounds are produced in specialized chromaffin cells and concentrated in membrane-bound vesicles of c. 50 μm diameter termed *chromaffin granules*. When the medulla is stimulated by the sympathetic nervous system, the catecholamines are released into the blood. Differential release of epinephrine and norepinephrine occurs under different physiological conditions and this, together with histochemical evidence, indicates that they are made and stored in different chromaffin cells. Epinephrine stimulates heart beat by inducing polarization changes at the pacemaker node, and also influences intermediary metabolism in most tissues. Norepinephrine is primarily involved in the contraction of smooth muscle.

Chromaffin granules can be isolated in quantity from bovine adrenal medulla, and resealed membrane ghosts can be prepared by hypoosmotic lysis. The catecholamines are highly concentrated within the vesicles, and are complexed with ATP and internal acidic proteins (chromogranin). The vesicles and chromogranin are probably derived from the ER-Golgi system, but the catecholamines enter directly from the cytosol by a carrier-mediated process that shows saturation kinetics and can be specifically inhibited. The granules have some dopamine-β-hydroxylase activity which suggests that dopamine taken up from the cytosol may be transformed to norepinephrine and epinephrine inside the granules. In recent years evidence has accumulated for a chemiosmotic mode of amine uptake. Data reviewed by Njus and Radda (1978) indicate that the membrane contains an ATPase that acts as a proton pump and generates a transmembrane pH difference

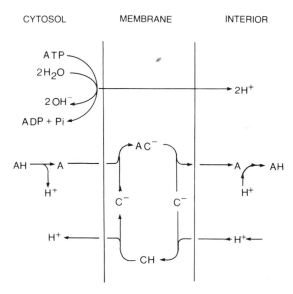

Figure 9.1   A model of catecholamine uptake into isolated chromaffin granules. A negatively charged carrier C^- binds on uncharged lipophilic amine and moves under the influence of the positive potential ($\Delta\psi$) to the inner face. Dissociation of the complex occurs there and the released amine becomes protonated. The carrier is also protonated and is driven to the outer face at the expense of the ΔpH.

(acid inside) and an electric potential (+ inside) giving a total proton motive force of about 140 mV (Johnson & Scarpa, 1979) that drives amine uptake (figure 9.1). Amine uptake is inhibited if the potential is reduced by addition of proton-conducting ionophores and permeant anions or by lack of ATP. This is an interesting finding that underlines chemiosmosis as a principle that extends beyond chloroplasts, mitochondria and bacteria.

Kostron *et al.* (1977) have shown that ATP accumulates in chromaffin granules independently of catecholamine uptake; the proton pump provides the energy, and uptake is mediated by a carrier that is sensitive to atractyloside, an inhibitor of the mitochondrial ATP/ADP carrier. Internal ATP is not used by the ATP-driven proton pump, which is primed by external ATP, and its role is undefined. Other unresolved questions include the roles of an NADH-acceptor oxidoreductase and cytochrome b-561 that are present in the membrane and may be involved in the transfer of reducing equivalents to dopamine-β-hydroxylase.

In vivo stimulation of the adrenal medulla by the splanchnic nerve induces catecholamine release. Experiments with perfused glands and isolated cells show that this is mediated by acetylcholine, the chemical produced at the nerve endings. Acetylcholine makes the plasma membranes permeable to external Ca^{2+} which enters and triggers exocytosis of the chromaffin granules.

9.2 Synaptic vesicles

In the neuromuscular system, the terminal contacts (synapses) between neurones are characterized by fine fluid-filled gaps of up to 20 nm wide, and there are similar but larger gaps between neurones and muscle. When an impulse reaches the end of a neurone, the signal is carried across the gap by chemical transmitters that are secreted into the cleft by the presynaptic neurone and interact with membrane receptors on the postsynaptic side. In the early 1950s, vesicles of 40–180 nm diameter were reported in presynaptic nerve endings and eventually shown to store neurotransmitters and release them into the cleft when the nerve was stimulated. Good vesicle preparations have become available only recently, and so their biochemical characterization has lagged behind that of chromaffin granules. There are two broad categories of vesicles and possibly others not yet defined:

1. *Adrenergic vesicles* from catecholamine-secreting neurones (adrenergic neurones). These contain catecholamines, ATP and chromogranin, and are very similar to chromaffin granules, except that epinephrine is usually replaced by norepinephrine or dopamine, the main catecholamine neurotransmitters.

2. *Cholinergic vesicles* from neurones that secrete acetylcholine as a neurotransmitter (cholinergic neurones). These are 40–70 nm in diameter and contain acetylcholine and ATP.

The membranes of synaptic vesicles and chromaffin granules have some similarities—they both show high lipid:protein ratios, have net negative surface charges, and share some antigenic characteristics. It is not yet known whether neurotransmitters are packaged into synaptic vesicles by chemiosmotic means as proposed for chromaffin granules. ATPase activity and ATP-dependent uptake of catecholamines into adrenergic vesicles have been reported, but further studies are necessary.

When an action potential reaches the presynaptic area of a neurone, it opens voltage-dependent Ca^{2+} channels in the membrane and the resulting Ca^{2+} influx induces fusion of synaptic vesicles with the presynaptic membrane and exocytosis of neurotransmitter into the cleft. However, there is also some non-vesicular release of acetylcholine by leakage from the cytosol across the membrane (see Tauc, 1977; Marchbanks, 1977), the significance of which is currently under debate. Some of the first evidence for exocytosis was an observation by Katz that miniature potentials appeared at motor end-plates, even in the absence of nerve stimulation, apparently as the result of random release of "packets" or "quanta" of acetylcholine from the presynaptic nerve endings. The amount of acetylcholine necessary to produce a minipotential was about the same as the content of a vesicle ($4-10 \times 10^4$ molecules) and considerably less than that required for maximum activity, implying that the normal passage of nerve impulses involved the discharge of many vesicles into the synaptic cleft. In fact, continuous stimulation decreases the number of vesicles in the terminals and Heuser *et al.* (1979) have shown a correlation between the number of quanta of acetylcholine discharged into the synaptic cleft and the number of exocytotic sites detectable by freeze-fracture electron microscopy.

In most nervous tissues there is a marked association of vesicles with electron-dense specializations in the presynaptic membrane. These projections of the membrane into the cytosol vary in different tissues and organisms, and may be spheres, rods, plaques or ribbons arranged singly or in groups (reviewed by Osborne, 1977), but in all cases they appear to sequester vesicles. Frog neuromuscular junctions are characterized by about 300 narrow transverse bars of electron-dense material, each with 50 or so sequestered vesicles arranged in two rows, one on each side of the bar. Figure 9.2 is a representation of a transverse section through an "active zone" as it is termed, and shows two rows of protein particles

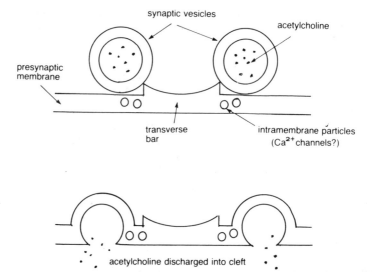

Figure 9.2   Diagrammatic representation of a transverse section through an active zone of a presynaptic membrane at the frog neuromuscular junction, showing neurotransmitter release into the cleft.

intrinsic in the membrane on each side of the bar under the vesicles. The vesicles are only 5 nm from the membrane. Heuser *et al.* (1979) have investigated exocytosis in this material, using a quick-freezing machine with a circuit that stimulates the junctions in the last few milliseconds before they are frozen. Analysis of freeze-fracture replicas showing vesicles caught in the act of exocytosis showed that the exocytotic sites were correlated with the active zones and not randomly distributed throughout the membrane. At maximum stimulation, 10 to 20 of the 50 vesicles per bar fused with the membrane, each event being independent of the others, i.e. there was no concerted action. Although the vesicle openings were randomly spaced along the active zone, they were regularly spaced relative to the rows of particles in the membrane, most of them being about 40 nm from the inner row (figure 9.2). Efflux of transmitter, its passage across the synaptic gap, and stimulation of the postsynaptic membrane all occur within 200 microseconds (μs) of the Ca^{2+} channels opening (Kelly *et al.*, 1979) and it has sometimes been argued that exocytosis is too slow to meet this timetable. In fact, the arrangements of vesicles only 5 nm from the membrane (possibly separated by an electrostatic or hydration barrier) means that they could be bathed in adequate Ca^{2+} concentrations within

20 μs. The proximity of the membrane particles to the vesicles is consistent with the view that they may provide Ca^{2+} channels directly to the vesicles, in which case Ca^{2+} activation would be even more rapid. According to Kelly *et al.* (1979), there is ample time for the active zones to be recharged with vesicles between impulses. Therefore it would appear that for this synapse at least, a satisfactory synthesis is emerging from neurophysiological, morphological and biochemical findings.

Exocytosis of synaptic vesicles must be accompanied by membrane recycling. One possibility is that membrane is recovered by coupled endocytosis of useful material released by exocytosis, e.g. adenine nucleotides and, in the case of adrenergic neurones, dopamine-β-hydrolase and chromogranin. Excess acetylcholine in the synaptic cleft is destroyed by acetylcholine esterase, but excess catecholamines are salvaged by the presynaptic neurone and, if this occurs by endocytosis, it may also contribute to membrane recovery. There have been reports of coated vesicles in nerve terminals shedding their coats and becoming synaptic vesicles; this encourages the view that receptor-mediated endocytosis could be fundamental to the recycling.

9.3 Sarcoplasmic reticulum

In skeletal muscle, each sarcomere of the myofibrils is enclosed in a sleeve of cisternae and tubules that is in intimate contact at regular intervals with the transverse tubules of the sarcolemma (figure 9.3). The sleeves of all the sarcomeres in the same cross striation form a continuum or network by means of interconnecting channels, and the whole is termed the sarcoplasmic reticulum (SR). SR is a Ca^{2+} reservoir and is present in all muscle systems with some type-specific variations. When the sarcolemma of a muscle fibre is excited by a nerve impulse, depolarization of the potential across the sarcolemma occurs and is communicated through the transverse tubules to the SR. The permeability of the SR membrane changes, and within a few milliseconds the Ca^{2+} floods out, bringing the concentration in the myoplasm surrounding the myofibrils up from about $0{\cdot}1\ \mu$M to 5–$10\ \mu$M. This triggers interaction between the myosin and actin filaments, and brings about ATP hydrolysis and muscle contraction. When the impulse passes, the SR becomes less permeable, Ca^{2+} is pumped rapidly back into the SR and muscle relaxation occurs, a process that takes between 10 and 100 ms, or rather longer in the case of slow muscles. SR therefore plays a fundamental role in the regulation of muscular activity by rapid Ca^{2+} release and removal, and is correspondingly adapted for this

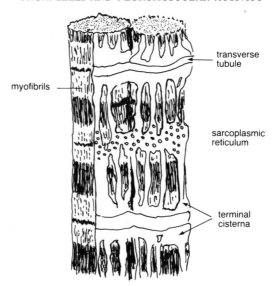

myofibrils

transverse
tubule

sarcoplasmic
reticulum

terminal
cisterna

Figure 9.3   A representation of the arrangement of sarcoplasmic reticulum relative to the
myofibrils of skeletal muscle.

function. Its proximity to the myofibrils means that the diffusion distance
for Ca^{2+} is less than 0.5 μm in fast muscle, and it has a vast surface area to
volume ratio calculated at about $10\,000$ cm^2 per g muscle. Even so,
membrane translocation of the order of 10^{12} Ca^{2+} ions cm^{-2}ms^{-1} must be
achieved to account for the observed kinetics—fast by any standard.
Uptake of Ca^{2+} is effected by a membrane-sited ATP-driven Ca^{2+} pump,
but little is known about the mechanism of Ca^{2+} efflux. A role for
mitochondria in Ca^{2+} uptake has frequently been proposed, but kinetic
considerations indicate that at best they could provide only back-up
facilities in the event of metabolic disturbances that cause high free Ca^{2+}
levels in the myoplasm.

Investigations into SR have been facilitated by the isolation of relatively
pure sarcoplasmic vesicles derived from SR by homogenization and
resealing. These vesicles rapidly take up Ca^{2+} in the presence of ATP,
particularly if a permanent anion like phosphate is there, and can reduce
external Ca^{2+} levels to below 0.1 μM. Two Ca^{2+} ions are translocated for
every ATP hydrolyzed:

$$2\ Ca^{2+}_{out} + ATP \ \rightleftharpoons\ 2\ Ca^{2+}_{in} + ADP + Pi$$

The Ca^{2+}-activated ATPase has been isolated and has a MW of $100\,000$;

it constitutes up to 90% of the intrinsic protein of the SR membranes. It is a transmembrane protein, and labelling experiments have shown that it has a large area exposed to the myoplasm phase and rather less on the lumen side. Two binding sites for Ca^{2+} and one for ATP have been demonstrated on the myoplasm side. ATP hydrolysis results in phosphorylation of the protein and in rotational or ionophoric translation of the Ca^{2+} to the inner surface. Experiments involving incorporation of enzymically cleaved fragments of the protein into liposomes suggest that part of the ATPase has ionophoric properties for Ca^{2+}, but this type of experiment must be interpreted cautiously. A low-MW proteolipid seems to interact functionally during the transport, as Ca^{2+} uptake into liposomes containing the ATPase is faster if the proteolipid is also incorporated in the membrane. An extrinsic high-affinity Ca^{2+} binding protein and several other proteins have also been isolated from SR. It will be some years before there is a consensus on kinetic models, far less their structural counterparts, for there is a constant flow of new information to be evaluated and assimilated, e.g. there is now evidence for a second phosphorylated state, a regulator site for ATP, and a role for K^+, cyclic AMP and protein kinases. SR contains a low-affinity high-capacity Ca^{2+}-binding protein (calsequestrin) which is sited mainly in the terminal cisternae near the transverse tubules. This protein binds 45 Ca^{2+} ions per molecule, and is probably important as an intra-SR sequestration site.

Studies on the biosynthesis of SR in differentiating muscle in tissue culture have recently been described by MacLennan and Campbell (1979). SR starts as outgrowths of the rER and the Ca^{2+}-dependent ATPase protein appears to be directly inserted into the membrane by bound ribosomes. In contrast, calsequestrin is detected in the rER-Golgi region first, and is probably synthesized into the lumen of rER and glycosylated before being packaged in the Golgi into vesicles that fuse with the developing SR or alternatively move through cisternae between rER and SR. Like mitochondria, ER and other organelles, SR shows a high degree of functionally relevant differentiation in different tissues; for example, Pette and Heilmann (1979) have reported that, compared with slow twitch muscles, SR from fast twitch muscles have (on an SR protein basis) ten times faster Ca^{2+}-dependent ATPase activity, ten times greater Ca^{2+} storage capacity, and thirty times faster uptake rate, as well as qualitative and quantitative differences in constituent proteins. Since SR is a relatively simple membrane with some well-defined marker proteins, it could well be the first internal membrane system to be properly characterized and understood in terms of the mechanics of biogenesis and control of differentiation.

BIBLIOGRAPHY

Chapter 1

FAWCETT, D. W. (1966) *The Cell: An Atlas of Fine Structure.* Saunders, Philadelphia.
GUNNING, B. E. S. & STEER, M. W. (1975) *Ultrastructure and the Biology of Plant Cells.* Edward Arnold, London.
GUNNING, B. E. S. & STEER, M. W. (1975) *Plant Cell Biology, an Ultrastructural Approach.* Edward Arnold, London.
HALL, J. L., FLOWERS, T. J. & ROBERTS, R. M. (1974) *Plant Cell Structure and Metabolism.* Longman.
SMITH, H. (1977) *The Molecular Biology of Plant Cells.* Botanical Monographs, Vol. 14. Blackwell Scientific Publications, Oxford.
STEPHENSON, W. K. (1979) *Concepts in Cell Biology.* Wiley, New York.

Chapter 2

ABELSON, J. (1979) RNA processing and the intervening sequence problem. *Ann. Rev. Biochem.* **48**, 1035–69.
ALOV, I. A. & LYUBSKII, S. L. (1977) Functional morphology of the kinetochore. *Int. Rev. Cytol. Suppl.* **6**, 59–74.
CAMPBELL, A. M. (1978) Straightening out the supercoil. *Trends in Biochemical Science* **3**, 104–108.
CAMPOUX, J. J. (1978) Proteins that affect DNA Conformation. *Ann. Rev. Biochem.* **47**, 449–479.
CHAMBON, P. (1975) Eukaryotic nuclear RNA polymerases. *Ann. Rev. Biochem.* **44**, 613–638.
ELGIN, S. C. R. & WEINTRAUB, H. (1975) Chromosomal proteins and chromatin structure. *Ann. Rev. Biochem.* **44**, 725–774.
GOIDL, J. A. & ALLEN, W. R. (1978) Does protein synthesis occur within the nucleus? *Trends in Biochemical Science* **3**, N225.
GOULD, R. R. & BORISY, G. G. (1977) The pericentriolar material in Chinese Hamster Ovary cells nucleates microtubule formation. *J. Cell Biol.* **73**, 601–615.
GRIFFITH, L. W. & POLLARD, T. D. (1978) Evidence for actin filament-microtubule interaction mediated by microtubule-associated proteins. *J. Cell Biol.* **78**, 958–965.
KORNBERG, R. D. (1977) Structure of chromatin. *Ann. Rev. Biochem* **46**, 931–954.
LILLEY, D. M. J. & PARDON, J. F. (1979) Chromatin and nucleosomes. *Chemistry in Britain* **15**, 182–190.
MACLEAN, N. (1976) *Control of Gene Expression.* Academic Press, London.
MAUL, G. G. (1977) The nuclear and cytoplasmic pore complex. *Int. Rev. Cytol. Suppl.* **6**, 76–186.
PAINE, P. L., MOORE, L. C. & HOROWITZ, S. B. (1975) Nuclear envelope permeability. *Nature* **254**, 109–114.

SHEININ, R., HUMBERT, J. & PEARLMAN R. E. (1978) Eukaryotic DNA replication. *Ann. Rev. Biochem.* **47**, 277–316.

SIEBERT, G. (1978) The limited contribution of the nuclear envelope to metabolic compartmentation. *Biochem. Soc. Trans.* **6**, 5–9.

SOEDA, E., MIURA, K., NAKASO, A. & KIMURA, G. (1977) Nucleotide sequence around the replication origin of polyoma virus DNA. *FEBS Letts.* **79**, 383–389.

TILGHMAN, S. M. *et al.* (1978) The intervening sequence of a mouse β-globin gene is transcribed within the 15S β-globin mRNA precursor. *Proc. Nat. Acad. Sci. U.S.A.* **75**, 1309–1313.

TIPPIT, D. H., SCHULZ, D. & PICKETT-HEAPS, J. D. (1978) Analysis of the distribution of spindle microtubules in the diatom *Fragilaria. J. Cell Biol.* **79**, 737–763.

WASSERMAN, W. J. & SMITH, L. D. (1978) The cyclic behaviour of a cytoplasmic factor controlling nuclear membrane breakdown. *J. Cell Biol.,* R15–R22.

WEINTRAUB, H., WORCEL, A. & ALBERTS, B. M. (1976) A model for chromatin based on two symmetrically paired half-nucleosomes. *Cell* **9**, 409–411.

WICKNER, S. H. (1978) DNA replication proteins of *E. coli. Ann. Rev. Biochem.* **47**, 1163–1191.

YAMAMOTO, K. R. & ALBERTS, B. M. (1976) Steroid receptors: elements for modulation of eukaryotic transcription. *Ann. Rev. Biochem.* **45**, 721.

Chapter 3

Further Reading

AKOYUNOGLOU, G. & ARGYROUDI-AKOYUNOGLOU, J. H. (1978) *Chloroplast development. Development in Plant Biology 2.* Elsevier/North Holland, Biomedical Press.

ARNTZEN, C. J. & BRIANTAIS, J. M. (1975) *Chloroplast structure and function.* In *Bioenergetics of Photosynthesis*, pp. 52–113 (ed. Govindjee). Academic Press, New York.

BARBER, J. (1976) *The intact chloroplast. Topics in Photosynthesis*—Volume 1. Elsevier/North Holland, Biomedical Press, Amsterdam, New York, Oxford.

BOARDMAN, N. K., ANDERSON, J. M. & GOODCHILD, D. J. (1978) *Chlorophyll-protein complexes and structure of mature and developing chloroplasts.* In *Current Topics in Bioenergetics*, Vol. 8, pp. 36–109. Academic Press, New York, San Francisco, London.

CALVIN, M. & BASSHAM, J. A. (1962) *The Photosynthesis of Carbon Compounds.* Benjamin, New York.

ELLIS, R. J. (1976) *Protein and nucleic acid synthesis by chloroplasts.* In *The Intact Chloroplast*, pp. 335–364. (ed. J. Barber). Elsevier/North Holland, Biomedical Press.

ELLIS, R. J., HIGHFIELD, P. E., SILVERTHORNE, J. (1978) The synthesis of chloroplast proteins by subcellular systems. In *Photosynthesis '77*, Proceedings of the Fourth International Congress on Photosynthesis 1977, pp. 497–506 (eds. D. O. Hall, J. Coombs, T. W. Goodwin). The Biochemical Society, London.

FORD, D. C. (1977) *Photosynthesis in the Science of Photobiology*, pp. 329–369. (ed. K. C. Smith) Plenum/Rosetta.

GOVINDJEE (1975) *Bioenergetics of Photosynthesis.* Academic Press, New York.

GREGORY, R. P. F. (1977) *Biochemistry of Photosynthesis.* 2nd ed. Wiley, London.

HALL, D. O., COOMBS, J. & GOODWIN, T. W. (1978) *Photosynthesis '77.* Proceedings of the Fourth International Congress on Photosynthesis. The Biochemical Society, London.

HEBER, U. (1974) Metabolite exchange between chloroplasts and cytoplasm. *Ann. Rev. Plant Physiol.* **25**, 393–423.

HELDT, H. W. (1976) *Metabolite transport in intact spinach chloroplasts.* In *The Intact Chloroplast*, pp. 215–234 (ed. J. Barber). Elsevier/North Holland Biomedical Press.

HEROLD, A. (1979) Nutrient transport in plant cells and tissues: transport of nutrients across chloroplast membranes. In *CRC Handbook*—Series in Nutrition & Food (in press) (ed. M. Recheigl). C.R.C. Press, Florida, U.S.A.

HINKLE, P. C. & McCARTY, R. E. (1978) How cells make ATP. *Sci. Amer.* March 1978, pp. 104–123.

JAGENDORF, A. T. (1975) *Mechanism of photophosphorylation.* In *Bioenergetics of Photosynthesis*, pp. 413–492 (ed. Govindjee). Academic Press, New York.

KIRK, J. T. O. & TILNEY-BASSETT, R. A. E. (1978) *The Plastids, their Chemistry, Structure, Growth and Inheritance.* Revised second edition. Elsevier/North Holland Biomedical Press.

LEA, P. J. & MIFLIN, B. J. (1974) Alternative route for nitrogen assimilation in higher plants. *Nature* **251**, 614–6.

LEECH, R. M. (1968) *The chloroplast inside and outside the cell.* In *Plant Cell Organelles*, pp. 137–159 (ed. J. B. Pridham). Academic Press, London, New York.

LEECH, R. M. & MURPHY, D. G. (1976) *The co-operative function of chloroplasts in the biosynthesis of small molecules.* In *The Intact Chloroplast*, pp. 365–402 (ed. J. Barber). Elsevier/North Holland Biomedical Press.

MIFLIN, B. J. & LEA, P. J. (1976) The pathway of nitrogen assimilation in plants. *Phytochem.* **15**, 873–85.

OLSON, J. M. & HIND, G. (1977) *Chlorophyll—Proteins, Reaction Centers and Photosynthetic Membranes.* Brookhaven Symposia in Biology No. 28.

PORTER, G. & ARCHER, M. D. (1976) *In vitro* photosynthesis. *Interdisciplinary Science Reviews* I (2), pp. 119–143.

SIEGELMAN, H. W. & HIND, G. (1978) *Photosynthetic Carbon Assimilation.* Basic Life Sciences Vol. II. Plenum Press, New York, London.

STAEHELIN, L. A., ARMOND, P. A. & MILLER, K. R. (1976) *Chloroplast membrane organisation at the supramolecular level and its functional implications.* In *Chlorophyll—Proteins, Reaction Centers and Photosynthetic Membranes.* Brookhaven Symposia in Biology No. 28.

THORNBER, J. P., ALBERTE, R. S., HUNTER, F. A., SHIOZAWA, J. A. & KAN, K-S (1977) *The organisation of chlorophyll in the plant photosynthetic unit.* In *Chlorophyll—Proteins, Reaction Centers, and Photosynthetic Membranes.* Brookhaven Symposia in Biology No. 28. pp. 132–148 (ed. J. M. Olson & G. Hind).

TREBST, A. (1974) Energy conservation in photosynthetic electron transport of chloroplasts. *Ann. Rev. Plant Physiol.* **25**, 423–58.

WALKER, D. A. (1974) *Chloroplast and cell—the movement of certain key substances etc. across the chloroplast envelope.* In *M.T.P. Int. Rev. of Science*, Biochemistry Ser. 1, Vol II: Plant Biochemistry, pp. 1–49 (ed. D. H. Northcote) Butterworth, London.

WALKER, D. A. & ROBINSON, S. P. (1978) *Regulation of photosynthetic carbon.* In *Photosynthetic Carbon Assimilation*, pp. 43–60 (eds. H. W. Siegelman & G. Hind). Plenum Press, New York, London.

Literature Cited

ANDERSON, M. (1975*a*) Possible location of chlorophyll within chloroplast membranes. *Nature* **253**, 536–7.

ANDERSON, J. M. (1975*b*) The molecular organisation of chloroplast thylakoids. *Biochem. Biophys. Acta.* **416**, 191–235.

BEALE, S. I., GOUGH, S. P. & GRANICK, S. (1975) Biosynthesis of δ-aminolevulinic acid from the intact carbon skeleton of glutamic acid in greening barley. *Proc. natn. Acad. Sci. U.S.A.* **72**, 2719–23.

BLAIR, G. E. & ELLIS, R. J. (1973) Protein Synthesis in chloroplasts. I. Light-driven synthesis of the large subunit of Fraction I protein by isolated pea chloroplasts. *Biochim. Biophys. Acta.* **319**, 223–4.

DOUCE, R. (1974) Site of biosynthesis of galactolipids in spinach chloroplasts. *Science N.Y.* **183**, 852–3.

GIVAN, C. & HARWOOD, J. (1976) Biosynthesis of small molecules in chloroplasts of higher plants. *Biol. Rev.* **51**, 365–406.

HELDT, H. W. & RAPLEY, L. (1970) Specific transport of inorganic phosphate, 3-phosphoglycerate and dihydroxyacetone phosphate, and of dicarboxylates across the inner membrane of spinach chloroplasts. *Fedn. Europ. Biochem. Soc. Lett.* **10**, 143–8.

HILL, R. & BENDALL, F. (1960) Function of the two cytochrome components in chloroplasts: a working hypothesis. *Nature* **186**, 136–7.

HIND, G. & JAGENDORF, A. T. (1963) Separation of light and dark stages in photophosphorylation. *Proc. Natl. Acad. Sci. U.S.A.* **49**, 715–22.

JENSEN, R. G. & BASSHAM, J. A. (1966) Photosynthesis by isolated chloroplasts. *Proc. Natn. Acad. Sci. U.S.A.* **56**, 1095.

LEECH, R. M. (1978) *Subcellular fractionation techniques in enzyme distribution studies.* In *Regulation of Enzyme Synthesis and Activiity in Higher Plants* pp. 290–324, (ed. Smith, H.). Academic Press.

LILLEY, R. McC. & WALKER, D. A. (1975) Carbon dioxide assimilation by leaves, isolated chloroplasts and ribulose carboxylase from spinach. *Plant Physiol.*, **55**, 1087.

LINK, G., COEN, D. M. & BOGORAD, L. (1978) Differential expression of the gene coding for the large subunit of ribulose-1,5 bisphosphate carboxylase in *Zea mays*. In *Chloroplast Development.* pp. 559–564 (ed. G. Akoyunoglou and J. H. Argyroudi-Akoyunoglou), Elsevier-North Holland.

RATHNAM, C. K. M. (1978) Malate and dihydroxyacetone phosphate-dependent nitrate reduction in spinach leaf protoplasts. *Plant Physiol.* **62**, 220–224.

STOCKING, C. R. & LARSON, S. (1969) A chloroplast cytoplasmic shuttle and the reduction of extraplastidic NAD. *Biochem. Biophys. Res. Commun.* **37**, 278–82.

STUMPF, P. K. (1975) Biosynthesis of fatty acids in spinach chloroplasts. In *Recent Advances in the Chemistry and Biochemistry of Plant Lipids*, pp. 95–113, (eds. T. Gaillard & E. I. Mercer), Academic Press, London.

Chapter 4

Further Reading

BOYER, P. D., CHANCE, B., ERNSTER, L., MITCHELL, P., RACKER, E. and SLATER, E. C. (1977) Oxidative phosphorylation and photophosphorylation. *Ann. Rev. Biochem.* **46**, 955–1026.

MITCHELL, P. (1976) Vectorial chemistry and the molecular mechanisms of chemiosmotic coupling: power transmission by proticity. *Biochem. Soc. Trans.* **4**, 399–430.

MOORMAN, A. F. M., VAN OMMEN, G-J. B. & GRIVELL, L. A. (1978) Transcription in yeast mitochondria: Isolation and physical mapping of messenger RNAs for subunits of cytochrome c oxidase and ATPase. *Molec. gen. Genet.* **160**, 13–24.

MUNN, E. A. (1974) *The Structure of Mitochondria.* Academic Press, London.

PALMER, J. M. (1976) The organisation and regulation of electron transport in plant mitochondria. *Ann. Rev. Plant Physiol.* **27**, 133–180.

ROTTENBERG, H. (1975) The measurement of transmembrane electrochemical proton gradients. *Bioenergetics* **7**, 61–74.

SLONIMSKI, P. P. & TZAGOLOFF, A. (1976) Localization in yeast mitochondria DNA of mutations expressed in a deficiency of cytochrome oxidase and/or coenzyme QH_2-cytochrome c reductase. *Eur. J. Biochem* **61**, 27–41.

Literature Cited

BRAND, M. D. (1979) Stoichiometry of charge and proton translocation in mitochondria. *Biochem. Soc. Trans.* (in press).

CHANCE, B. (1972) The nature of electron transfer and energy coupling reactions. *FEBS Letts.* **23**, 3–20.

COTE, C., SOLIOZ, M., SCHATZ, G. (1979) Biogenesis of the Cytochrome bc_1 complex of Yeast Mitochondria. *J. Biol. Chem.* **254**, 1437–1439.

HINKLE, P. C. & WU, M. L. (1979) The phosphorus/oxygen ratio of mitochondrial oxidative phosphorylation. *J. Biol. Chem.* **254**, 2450–2455.

LUCK, D. J. L. (1963) Formation of mitochondria in *Neurospora crassa. J. Cell Biol.* **16**, 483–499.

MACCECCHINI, M-L., RUDIN, Y., BLOBEL, G. & SCHATZ, G. (1979) Import of proteins into mitochondria. *Proc. Nat. Acad. Sci. U.S.A.* **76**, 343–347.

MACINO, G. & TZAGOLOFF, A. (1979) The DNA sequence of a mitochondrial ATPase gene in *Saccharomyces cerevisiae. J. Biol. Chem.* **254**, 4617–4623.

MAHLER, H. R. (1973) Biogenetic autonomy of mitochondria. *CRC Critical Reviews in Biochemistry*, pp. 381–460.

MAYER, R. J. (1979) Turnover of mitochondrial proteins in rat liver. *Biochem. Soc. Trans.* **7**, 306–310.

MITCHELL, P. (1961) Coupling of phosphorylation to electron and hydrogen transfer by a chemi-osmotic type of mechanism. *Nature* **191**, 144–147.

MITCHELL, P. (1979) Compartmentation and communication in living systems. Ligand conduction: a general catalytic principle in chemical, osmotic and chemiosmotic reaction systems. *Eur. J. Biochem.* **95**, 1–29.

MITCHELL, P. & MOYLE, J. (1979) Respiratory-chain protonmotive stoichiometry. *Biochem. Soc. Trans.* (in press).

NASS, M. M. K. & NASS, S. (1963) Intramitochondrial fibres with DNA characteristics. *J. Cell Biol.* **19**, 613–629.

PALMER, J. M. (1979) The 'uniqueness' of plant mitochondria. *Biochem. Soc. Trans.* **7**, 246–252.

RACKER, E. (1975) Reconstitution, mechanism of action and control of ion pumps. *Biochem. Soc. Trans.* **3**, 785–802.

REID, R. A. (1974) Reversal of adenosine triphosphatase in chloroplasts and mitochondria by transmembrane ion gradients. *Biochem. Soc. Spec. Publ.* **4**, 113–129.

SLONIMSKI, P. P. *et al.* (1978) *Mosaic organization and expression of the mitochondrial DNA region controlling cytochrome c reductase and oxidase.* In *Biochemistry and Genetics of Yeast*, ed. Bacila, M., Horecker, B. L. & Stoppani, A. D. M., Academic Press.

SMITH, G. S. & REID R. A. (1978) The influence of respiratory state on monoamine oxidase activity in rat liver mitochondria. *Biochem. J.* **176**, 1011–1014.

TZAGOLOFF, A., MACINO, G. & SEBALD, W. (1979) Mitochondrial genes and translation products. *Ann. Rev. Biochem.* **48**, 419–441.

Chapter 5

BEEVERS, H. (1975) *Organelles from castor bean seedlings: biochemical roles in gluconeogenesis and phospholipid biosynthesis.* In *Recent Advances in Chemistry and Biochemistry of Plant Lipids*, pp. 287–299, (eds. T. Galliard & E. I. Mercer). Academic Press, New York.

BREIDENBACH, R. W. & BEEVERS, H. (1967) Association of the glyoxylate cycle enzymes in a novel subcellular particle from castor bean endosperm. *Biochem. Biophys. Res. Comm.* **27**, 462–469.

COOPER, T. G. & BEEVERS, H. (1969) Mitochondria and glyoxysomes from castor bean endosperm. *J. Biol. Chem.* **244**, 3507–3513.

COOPER, T. G. & BEEVERS, H. (1969) β-oxidation in glyoxysomes from castor bean endosperm. *J. Biol. Chem.* **244**, 3514–3520.

HUANG, A. H. C., BOWMAN, P. D. & BEEVERS, H. (1974) Immunological and biochemical studies on isozymes of malate dehydrogenese and citrate synthetase in castor bean glyoxysomes. *Plant Physiol.*, **54**, 364–8.

MASTERS, C. & HOLMES, R. (1977) The metabolic roles of peroxisomes in mammalian tissues. *Int. J. Biochem.* **8**, 549–559.

TOLBERT, N. E. (1971) Microbodies-peroxisomes and glyoxysomes. *Ann. Rev. Plant Physiol.* **22**, 45–197.

Chapter 6

Further Reading

BOWLES, D. J. & NORTHCOTE, D. H. (1976) The size and distribution of polysaccharides during their synthesis within the membrane system of maize root cells. *Planta* (Berl.) **128**, 101–6.

BROWN, R. M., FRANKE, W. W., KLEINIG, H., FALK, H. & SITTE, P. (1970) Scale formation in chrysophycean algae. *J. Cell Biol.* **45**, 246.

HOWELL, K. E., ITO, A. & PALADE, G. L. (1978) Endoplasmic reticulum marker enzymes in Golgi fractions—what does this mean? *J. Cell Biol.* **79**, 581–589.

LEADER, D. P. (1979) Protein synthesis on membrane-bound ribosomes. *Trends in Biochem. Sc.* **4**, 205–208.

MOLLENHAUER, H. H. & MORRÉ, J. D. (1966) Golgi apparatus and plant secretions. *Ann. Rev. Plant Physiol.* **18**, 27.

MORRÉ, J. D., MOLLENHAUER, H. H. & BRACKER, C. E. (1970) *Origin and continuity of Golgi apparatus.* In *Origin and Continuity of Cell Organelles*, (eds. J. Reinert & H. Ursprung). Springer Verlag, Berlin.

NORTHCOTE, D. H. (1968) *The organisation of the endoplasmic reticulum, the Golgi bodies and microtubules during cell division and subsequent growth.* In *Plant Cell Organelles*, pp. 179–197, (ed. J. B. Pridham). Academic Press, London.

WHALEY, W. G. (1975) *The Golgi Apparatus.* (Cell Biology Monograph No. 2). Springer Verlag, Wien.

Literature Cited

DALLNER, G., ELHAMMER, A. & VALTERSSON, C. (1979) Biosynthesis and assembly of the endoplasmic reticulum. *Biochem. Soc. Trans.* **7**, 297–300.

DASHEK, W. V. & ROSEN, W. G. (1966) Electron microscopical localisation of chemical components in the growth zone of lily pollen tubes. *Protoplasma* **61**, 192.

GRATZL, M. & DAHL, G. (1978) Fusion of secretory vesicles isolated from rat liver. *J. Membrane Biol.* **40**, 343–364.

KELLY, R. B., DEUTSCH, J. W., CARLSON, S. S. & WAGNER, J. A. (1979) Biochemistry of neurotransmitter release. *Ann. Rev. Neuroscience* **2**, 399–446.

NORTHCOTE, D. H. & PICKETT-HEAPS, J. D. (1966) A function of the Golgi apparatus in polysaccharide synthesis and transport in the root-cap cells of wheat. *Biochem. J.* **98**, 159–167.

PALADE, G. L. (1975) Intracellular aspects of the process of protein secretion. *Science* **189**, 347–359.

PAPAHADJOPOULOS, D., VAIL, W. J., NEWTON, C., NIR, S., JACOBSON, K., POSTE, G. & LAZO, R. (1977). Studies on membrane fusion. III The role of calcium-induced phase changes. *Biochim. Biophys. Acta* **465**, 579–598.

SABATINI, D. D. & BLOBEL, G. (1970) Controlled proteolysis of nascent polypeptides in rat liver cell fractions II. Location of the polypeptides in Rough Microsomes. *J. Cell Biol.* **45**, 146–157.

VON HEIJNE, G. & BLOMBERG, C. (1979) Trans-membrane translocation of proteins: the direct transfer model. *Eur. J. Biochem.* **97**, 175–181.

170 BIBLIOGRAPHY

Chapter 7

Further Reading

DEAN, R. T. (1977) Lysosomes and membrane recycling. *Biochem. J.* **168**, 603–605.
HALL, J. L., FLOWERS, T. J. & ROBERTS, R. M. (1974) *Plant Cell Structure and Metabolism.* Ch. 12 Lysosomes. Longman.
HOLTZMAN, E. (1975) *Lysosomes: a Survey.* (Cell Biology Monographs No. 3) Springer Verlag, Wien.
NEUFIELD, E., LIM, T. & SHAPIRO, L. (1975) Inherited disorders of lysosomal metabolism. *Ann. Rev. Biochem.* **44**, 357–376.

Literature Cited

DEAN, R. T. & BARRETT, A. J. (1976) Lysosomes. *Essays in Biochemistry* **12**, 1–40.
DE DUVE, C. (1959) *Lysosomes, a new group of cytoplasmic particles.* In *Subcellular Particles,* (ed. T. Hayashi) 129–159. Ronald Press, New York.
DINGLE, J. T. (1973) *Lysosomal enzymes in skeletal tissues.* In *Hard Tissue Growth, Repair and Remineralisation* (Ciba Foundation Symp. 11), pp. 295–313.
GOLDSTEIN, J. L., ANDERSON, R. G. W. & BROWN, M. S. (1979) Coated pits, coated vesicles and receptor mediated endocytosis. *Nature* **279**, 679–684.
GOLDSTEIN, J. L. & BROWN, M. S. (1977) The low density lipoprotein pathway and its relation to atherosclerosis. *Ann. Rev. Biochem.* **46**, 897–930.
NOVIKOFF, A. B. (1973) In *Lysosomes and Storage Diseases* (eds. Hers, H. G., Van Hoof, F.). Academic Press, New York and London.
NOVIKOFF, A. B. (1974) In *The Cytopharmacology of Secretion,* Advan. Cytopharmacol. 2 (eds. Ceccarrelli, B., Clementi, F. & Meldolesi, J.). Raven Press, New York.
REIJNGOUD, D.-J. (1978) The pH and transport of protons in lysosomes. Trends in Biochem. *Science* **3**, 178–180.

Chapter 8

Further Reading

HEBER, U. (1974) Metabolite exchange between chloroplasts and cytoplasm. *Ann. Rev. Plant Physiol.* **25**, 393–421.
HELDT, H. W., CHON, C. J., MARONDE, D., HEROLD, A., STANKOVIC, Z. S., WALKER, D. A., KRAMINER, A., KIRK, M. R., &HEBER, U. (1977) Role of Orthophosphate and other factors in the regulation of starch formation in leaves and isolated chloroplasts. *Plant Physiol.* **59**, 1146–55.
HELDT, H. W., FLIEGE, R., LEHNER, K., MILOVANCEV, M. & WERDAN, K. (1974) Metabolite movement and CO_2 fixation in spinach chloroplasts. In *Proc. IIIrd Int. Cong. on Photosynthesis,* pp. 1369–1380 (ed. M. Avron), Elsevier, Amsterdam.
KRAUSE, G. H. & HEBER, U. (1976) *Energetics of intact chloroplasts.* In *The Intact Chloroplast* pp. 171–214 (ed. J. Barber). Elsevier/North Holland Biomedical Press.
KUNG, S. (1977) The expression of chloroplast genomes in higher plants. *Ann. Rev. Plant Physiol.* **28**, 401–39.
MARGULIS, L. (1975) *Symbiotic theory for the origin of eukaryotic organelles: criteria for proof.* In *Symbiosis* (eds. Jennings, D. H. & Lee, D. L.), SEB Symposium **29**, 21–38.
NEWSHOLME, E. A. & START, C. (1973) *Regulation in Metabolism.* Wiley, London.
WALKER, D. A. (1976) CO_2 *fixation by intact chloroplasts: photosynthetic induction and its*

relation to transport phenomena and control mechanisms. In *The Intact Chloroplast*, pp. 235–278 (ed. J. Barber). Elsevier/North Holland Biomedical Press.

WERDAN, K., HELDT, H. W. & MILANOVANCEV, M. (1975) The role of pH in the regulation of CO_2 fixation in the chloroplast stroma. Studies in CO_2 fixation in light and dark. *Biochem. Biophys. Acta.* **396**, 276–92.

Literature Cited

BRODA, E. (1975) *The Evolution of the Bioenergetic Processes.* Pergamon Press, Oxford.

FUGE, H. (1977) Ultrastructure of the mitotic spindles. *Int. Rev. Cytol. Suppl.* **6**, 1–52.

KUBAI, D. F. (1975) The evolution of the mitotic spindle. *Int. Rev. Cytol.* **43**, 167–

KUNG, S. D. & RHODES, P. R. (1978) *Interaction of chloroplast and nuclear genomes in regulating RuBP carboxylase activity.* In *Photosynthetic Carbon Assimilation*, pp. 307–324 (eds. H. W. Siegelman & G. Hind) Plenum Press, New York, London.

Chapter 9

INGEBRETSEN, O. C. & FLATMARK, T. (1979) Active and passive transport of dopamine into chromaffin granule ghosts isolated from bovine adrenal medulla. *J. Biol. Chem.* **254**, 3833–3837.

KOSTRON, H., WINKLER, H., PEER, L. J. & KONIG, P. (1977) Uptake of ATP by isolated adrenal chromaffin granules: A carrier-mediated transport. *Neuroscience* **2**, 159–166.

JOHNSON, R. G. & SCARPA, A. (1979) Protonmotive force and catecholamine transport in silated chromaffin granules. *J. Biol. Chem.* **254**, 3750–3760.

NJUS, D. & RADDA, G. K. (1978) Bioenergetic processes in chromaffin granules. A new perspective on some old problems. *Biochim. Biophys. Acta.* **463**, 219–244.

HEUSER, J. E., REESE, T. S., DENNIS, M. J., JAN, Y., JAN, L. & EVANS, L. (1979) Synaptic vesicle exocytosis captured by quick freezing and correlated with quantol transmitter release. *J. Cell Biol.* **81**, 275–300.

KATZ, B. (1969) *The Release of Neural Transmitter Substance.* Liverpool University Press, Liverpool.

KELLY, R. B., DEUTSCH, J. W., CARLSON, S. S. & WAGNER, J. A. (1979) Biochemistry of neurotransmitter release. *Ann. Rev. Neuroscience* **2**, 399–446.

OSBORNE, M. P. (1977) *Role of vesicles with some observations on vertebrate sensory cells.* In *Synapses* (Eds. Cottrell, G. A. & Usherwood, P. N. R.). pp. 40–63. Blackie, Glasgow & London.

MARCHBANKS, R. M. (1977) *Turnover and release of acetylcholine.* In *Synapses*, pp. 81–102 (eds. Cottrell, G. A. & Usherwood, P. N. R.) Blackie, Glasgow and London.

STARKE, K. (1977) Regulation of noradrenaline release by presynaptic receptor systems. *Rev. Physiol. Biochem. Pharmacol.* **77**, 1–124.

TAUC, L. (1977) *Transmitter release at cholinergic synapses.* In *Synapses*, pp. 64–80 (Eds. Cottrell, G. A. & Usherwood, P. N. R.) Blackie, Glasgow and London.

WEINER, N. (1979) Multiple factors regulating the release of norepinephrine consequent to nerve stimulation. *Fed. Proc.* **38**, 2193–2202.

HASSELBACH, W. (1978) The reversibility of the sarcoplasmic calcium pump. *Biochim. Biophys. Acta.* **515**, 23–53.

MACLENNAN, D. H. & CAMPBELL, K. P. (1979) Structure, function and synthesis of sarcoplasmic reticulum proteins. *Trends in Biochem. Sc.* **4**, 148–150.

PETTE, D. & HEILMANN, C. (1979) Some characteristics of sarcoplasmic reticulum in fast- and slow-twitch muscles. *Biochem. Soc. Trans.* **7**, 765–767.

Index

172